The Economics of Cable Television

The Economics of Cable Television

G. Kent Webb

LexingtonBooks
D.C. Heath and Company
Lexington, Massachusetts
Toronto

Library of Congress Cataloging in Publication Data

Webb, G. Kent.
The economics of cable television.

Bibliography: p.
Includes index.
1. Cable television—United States. I. Title.
HE8700.7.C6W4 1983 384.55'568'0973 82-48574
ISBN 0-669-06323-1

Published simultaneously in Canada

Printed in the United States of America

International Standard Book Number: 0-669-06323-1

Library of Congress Catalog Card Number: 82-48574

For Amber

Contents

Figures

Tables

Preface and Acknowledgments

Cable television began and continues to operate in many communities as nothing more elaborate than a good television antenna that can be shared among a number of households. In other communities, however, cable television has developed into a high-technology component of the communications revolution. The public responsibility for this development is local; most cable-television systems are privately owned but municipally franchised.

This book identifies the market and regulatory forces that have shaped the cable-television industry. More important, it presents an investigation of the issues of price, design, and regulation. These issues are important to corporate planners who are responsible for the billions of dollars invested in cable each year, to municipal authorities who have regulatory jurisdiction over cable, and to anyone who may sometime want to use a cable system.

Based on an econometric and engineering analysis of the market and technology, a computer model was designed to simulate the economic behavior of a cable-television system. The results of the simulation are described in general terms, with technical details confined to the appendixes. The results help explain why cable grew so quickly during the late 1970s and early 1980s, why it has begun to slow down, and where its going from here. Perhaps the most controversial conclusion is that cable television is a natural geographic monopoly and should be subject to some form of price regulation.

Among the many individuals who contributed ideas and encouragement to this project, I would particularly like to acknowledge Almarin Phillips, Robert de Vos, and Deborah Denno. I am also indebted to Philip A. Viton, Barry Cole, David Crawford, Bob Douglas, and Sharon Briley for their comments and technical guidance, and to students at both the University of Pennsylvania and Drexel University who assisted in collecting data. Erica Ginsburg deserves special thanks for drawing the figures that appear throughout the book.

1 Introduction

In 1948 an enterprising appliance dealer, John Watson, confronted the problem of selling television sets to the people of Mahony City, Pennsylvania. Reception in the area was poor because mountainous terrain blocked signals broadcast from Philadelphia. Watson's idea was to mount a television antenna on top of nearby Tuscarora Mountain. A long copper cable allowed townspeople to connect their sets to the antenna, providing a clear television picture. Once good reception was available, television sales boomed; Watson profited; and—so the story goes—cable television was born.

By the late 1970s cable systems were proposed for every major city in the country. By 1980 annual investment exceeded $1 billion. In recent years annual revenue growth has surpassed 20 percent. Emerging from the rural United States, cable television has become one of the growth industries of the decade. In the process a number of economic questions have been raised. Should cable television be regulated, should it be municipally owned, or will the open market best provide service? What are reasonable prices? What is the ideal system design?

Technological progress and market dynamics continually modify the answers to some of these questions. Uncertainty limits the conclusions. Even so, billions will be invested on the basis of the best available information.

This notion of best information has been expressed mathematically in the development of econometrics, the application of statistical techniques to economic problems. The useful products of econometrics include a measure of the uncertainty associated with available information and a method for constructing computer-based economic models of a firm. Such a model provides the basis for the economic investigation of cable television that is reported in this book. The details of model construction are included in appendixes.

To put the question at issue into perspective, chapters 1 and 2 are devoted to a description of telecommunications services and a history of the development of the industry structure. Chapter 3 focuses on the cost of operating a cable system; it includes a description of equipment and programming sources. The market for cable services is analyzed in chapter 4. Against this background and using Philadelphia as a case study, the remaining chapters investigate questions of price, design, and regulation.

Technical Overview

In its elementary form, a cable system consists of a well-placed television antenna that is connected to the sets of a number of subscribers. The system provides improved reception of local broadcast signals. More elaborate receiving equipment can be used to import signals from outside the local television market.

Signals received by cable systems usually require amplification and processing before they can be distributed to subscribers. The equipment used for this purpose is located at the headend, which houses the electronic parallel of a brain for the system. Among systems offering sophisticated telecommunications services, elaborate headend equipment is required. For example, newly constructed systems must be capable of two-way communication so that signals may travel upstream (from subscriber to headend) as well as downstream (from headend to subscriber). Extensive switching and computing equipment may be needed at the headend to manage and respond to this two-direction signal traffic.

Carrying signals to and from the subscriber is the function of the cable-system distribution plant, which is often characterized as an electronic pipeline. The resemblance between the distribution plant and a pipeline is both technical and economic. Like water or oil pumped through a pipeline, the signals of a cable system are pumped—*amplified* is the proper electronic description—through the cable distribution plant. Nearly all systems use coaxial cable, consisting of a copper core sheathed in plastic and aluminum to prevent interference. Cable made of glass fiber has been used for long-haul transmission of signals and will likely be used as the distribution plant of cable systems in the future. A few technical problems, discussed in chapter 3, still prevent widespread use of glass fiber.

The transmission capacity of the cable distribution plant is described in terms of the number of television channels it can carry. Through the early 1980s the industry standard has been 40-channel coaxial cable. Interference from local broadcast signals, may reduce the effective channel capacity, however. New systems have been designed using thicker, more expensive cable that can carry up to 54 channels. Increasing the channel capacity above the 54-channel range currently requires that more than one coaxial cable be installed in the system. For example, urban cable systems have been proposed that would require three or more cables and would provide more than 100 channels of programming. Additional channels may be dedicated to institutional use, allowing for video and data transmission by municipal authorities or by private firms.

An important factor in the cost of the distribution plant is the type of construction required for installation: aerial or underground. Less expensive, aerial construction simply involves attaching the cable to existing utility poles or houses. Underground installation is more expensive, especially when it requires the digging of trenches and the subsequent restoration of an urban landscape. When the cable can be run through existing underground utility conduits, the cost of underground cable is substantially reduced. Although expensive to install and maintain, underground cable is less obtrusive than aerial cable.

The equipment installed at the subscriber's end of the distribution plant serves as the interface between the cable system and the television set. A component of the interface for systems with two-way capability is the keypad, a microcomputer similar to an electronic calculator, which allows subscribers to send signals upstream to the headend. Using the keypad, subscribers to the more elaborate systems can purchase groceries, participate in public meetings, or vote on the best ending to a melodrama. Other devices—a smoke detector, appliance monitor, or computer terminal—may be connected to the system as part of the new interactive services being developed for the industry.

Most large cable systems devote several channels to local programming. The equipment required for local production in the studio of a cable system is essentially the same as might be found in the studio of a broadcast television station. It includes video cameras, recorders, and audio and lighting equipment. The extent of investment in these facilities is often a matter of controversy when franchise agreements are negotiated. The municipal authorities who grant the franchises want to encourage local programming and have increasingly emphasized extensive investment in local production facilities as a term of the franchise agreement. The cable operators who finance the systems are less eager to invest in local production because, while it is good public relations, local production is often unprofitable.

Historical Overview

Early cable systems, often called CATV (for *community antenna television*), were constructed where terrain or distance prohibited adequate reception of broadcast television. Most systems were built in the eastern United States. In the West, communities often chose a different technology, the television translator. Like a cable system, the translator uses an antenna mounted in a good location to receive distant broadcast signals. The television translator, however, acts as a relay broadcast system. It rebroadcasts the signals on new frequencies, boosting their power. An advantage which has favored the widespread development of cable is that, while the number of channels that can be distributed by a television translator is limited by the availability of broadcast frequencies, the channel capacity of a cable system is technologically unlimited.

At first the relationship between cable and broadcast television was harmonious. Cable systems increased the audience of broadcast stations by improving signal reception in distant parts of the local market. The broadcasters were pleased because a larger audience usually means more advertising revenue. The cable operators were pleased because broadcast signals provided a free source of programming. At the time, cable operators were under no obligation to pay for use of broadcast signals.

The harmony ended when cable operators began using microwaves to import distant television signals for distribution in communities that already had local broadcast stations. Previously, distance had prevented competition between stations in different markets. Now, conceivably, the local cable operator could carry the signals of broadcast stations from all over the country. Many broadcasters believed that this competition would lead to the development of a few large cable-distributed stations attracting a national audience at the expense of local broadcast stations. The prospect brought about the first national confrontation between cable and broadcast television. Broadcasters complained to the Federal Communications Commission (FCC).

The FCC, which had allocated the broadcast spectrum to encourage the development of local television, imposed regulatory constraints on cable to prevent the feared economic ruin of local broadcast stations. From 1962 until about 1977, the cable industry was tightly regulated. By the end of the 1970s, however, most of the FCC restraints were eliminated. The details of these crucial regulatory decisions appear in chapter 2. The results of these decisions are evident in the cable construction boom, which has coincided with the relaxation of federal regulations. Of course, not all the recent expansion can be attributed to federal deregulation. Cable television offers more than just a means of redistributing broadcast signals.

In 1972—about the time that marks the beginning of federal deregulation—Time, Inc., launched the first pay-television network, Home Box Office (HBO). Network programming on HBO includes recently released motion pictures, sports, and entertainment specials—all without commercial interruption. The central concept of pay television is that subscription fees rather than advertising sales are used to finance the network's programming. The average monthly subscription fee is about $10 but can range from $2 to $25 depending on the type of programming, local market conditions, and the technology used to distribute pay-television programs. About 90 percent of all pay-television programming is distributed to subscribers by cable systems; broadcast technologies account for the remaining market share.

At first, pay television networks, like cable television systems, had to weather restrictive FCC regulations designed to protect the advertising-supported broadcast industry. After the restrictions were eliminated, pay-television revenues contributed substantially to growth in the cable industry.

During the same period in the early 1970s that ushered in pay television, cable operators began experimenting with two-way systems. A two-way cable system, like a telephone system, allows subscribers both to send and to receive signals. The advantage of cable over current telephone systems is its larger transmission capacity, or bandwidth. For example, cable systems can transmit video, whereas conventional telephone systems cannot.

In 1972, Teleprompter and Hughes Aircraft developed and installed a two-way cable system but found that demand for its services did not justify the cost. Time, Inc., through its American Television and Communications (ATC) subsidiary, evaluated but decided not to construct a similar system for Orlando, Florida. A few years later, in Columbus, Ohio, Warner-Amex built its Qube system, which has been used to test-market a variety of two-way, or interactive, services. More recently, Qube has been installed in newly constructed cable systems in other parts of the country. Although much of the industry's future growth is predicated on the development of two-way services, a positive bottom-line impact from these innovations has been slow to come.

Innovations in satellite technology, however, brought rapid changes. RCA launched the first domestic satellite in 1975, and the sky has since become crowded with communications satellites. Among technologies that can be used to transmit microwaves, the cost advantage is with satellites when the distance is large and the receivers are geographically scattered. Cable operators throughout the country can focus their receiving equipment, called an earth station, on an individual satellite. Because microwaves require a line-of-sight transmission path, land-based systems are more restricted. The curvature of the earth and obstructions from the local terrain usually prevent land-based transmissions from traveling more than twenty-five or thirty miles.

The sudden increase in the availability of satellite channel capacity has allowed for programming tailored to specific audiences. *Narrow-casting,* as it is called, includes channels dedicated to sports, news, ethnic, and childrens' programming. Evangelists were quick to use the new medium to offer religious programming. Existing pay-television networks used the satellites to expand their markets; new networks were formed. In Atlanta, Georgia, the yachtsman and entrepreneur Ted Turner decided to distribute his WTBS-TV station to forty-eight states by satellite. Eventually he launched the Cable News Networks.

Cable-Television Service

Historically, the entertainment services offered by a cable system have been divided into two general categories: basic and pay cable. Recently this distinction has become somewhat muddled. *Basic cable* service includes the improved reception of local broadcast television, signals from distant stations, and low-cost satellite-distributed signals for which the subscriber is not directly billed. Basic cable is sometimes called regular subscriber service.

Pay cable refers to the cable distribution of pay-television network programming, such as offered by HBO. Subscribers are directly billed for pay-cable service, in much the same way that long-distance telephone service is billed separately from local telephone service. On most systems the optional pay-cable services are available only to households subscribing to the basic-cable system. Pay cable is sometimes called deluxe service; subscriber fees are largely used to pay for programming. Basic cable fees are largely devoted to the cost of building and operating the cable system.

Basic-Cable Service

The most common use of basic-cable channels is to distribute the signals of broadcast stations operating in the local market. Often, stations imported by microwave from neighboring television markets are also distributed on the cable systems. A few nationally distributed stations have emerged, as was feared by broadcasters when cable operators first imported signals by microwave. These superstations, as they have been called, offer a programming mix popular among stations not affiliated with ABC, CBS, NBC, or PBS. The programming includes sports, old movies, news, and syndicated programs such as "I Love Lucy," "Star Trek," and "Leave It to Beaver." Four stations had achieved superstation status by 1983: WTBS, Atlanta; WGN, Chicago; KTUV, San Francisco; and WOR, New York. These stations, which rely primarily on advertising revenues for financial support, can now be viewed by cable subscribers throughout the country.

In general, advertising rates—the price of a slot of time on the broadcast station—are directly related to the number of viewers watching the station. Accordingly, superstations raised their advertising rates as their audience grew. The drawback to superstation status has been that many firms that have traditionally advertised on local broadcast stations are interested only in reaching local viewers. For example, an Atlanta car dealer would be reluctant to pay higher advertising rates for the privilege of reaching a viewer in Billings, Montana tuned into the Atlanta station via cable.

The higher rates charged for advertising time by superstations have encouraged firms with local advertising goals to do business with a conventional broadcast station offering lower rates and serving only the local market. The potential clients of superstations are firms with national target audiences. CocaCola, General Foods, and Revlon, for example, have used the major television networks to reach national audiences. Superstations have risen above the competition of local markets to face direct competition with the national television networks. Consequently, fewer stations than had once been feared have actually sought superstation status and the associated competition.

A few systems include cable radio as a basic-cable service. Like basic-cable television, cable radio offers improved reception of local signals and some signals from distant sources. To the devotee, the signal quality offered by cable radio represents an appreciable breakthrough. Cable's broadband capacity allows for the distribution of high and low tone ranges that are chopped off at the extreme ranges of the FM broadcast signal. Cable-radio superstations, such as WFMT, Chicago, are nationally distributed to systems via satellite. Several new basic-cable services, such as The Music Channel, offer music combined with video images.

Most of the recent expansion in the number of basic-cable channels available to subscribers has been a result of new networks distributed to cable systems by satellite. Table 1-1 is a list of satellite-distributed networks with a description of the type of programming offered by each. Notice that in many instances, programming is targeted to a specific market segment.

Three sources of revenue have so far supported satellite-distributed basic-cable services: a monthly charge to the system operator, advertising revenues, and viewer donations. The charge to the system operator usually ranges from $0.01 to $0.20 per month per subscriber.

Many of the basic-cable networks delivering religious programming rely solely on viewer donations for support. Sales of items of religious interest through the network are often another source of revenue. This arrangement has been adapted successfullly from the broadcast industry, in which religious programs have for years been syndicated nationwide by radio and television.

Programming produced by the local cable operator is often partially supported by advertising. These programs, featuring local personalities, include coverage of area sports, news, and civic events. Many communities encourage public access to the system by requiring that cable channels and production equipment be available for community use. Citizens, citizen groups, or education and governmental institutions can produce their own programming for local distribution using these public-access channels. There are more than a thousand public-access channels among cable systems nationwide. Many industry participants believe that public access and locally produced programming represent the most important aspect of the industry: a video medium at the community level.

Another access concept, that of *leased channels,* has led to speculation about how the industry could be developed. Unlike public-access channels dedicated to community use, which are usually available at little or no cost, leased channels are available at commercial rates. For years, telephone lines

Table 1-1
Satellite-Distributed Basic-Cable Services, 1982

Service	*Programming*	*Start-up Date*
ACSN (Appalachian Community Service Network)	Educational and community service; emphasis is on college credit, continuing education, and professional-development programming.	8/79
ARTS (Alpha Repertory Television Service)	Visual and fine-arts programming featuring music, drama, and dance specials.	4/81
BET (Black Entertainment Television)	Programming oriented toward the black audience—feature films, musical specials, sports and public affairs.	3/79
CBN (Christian Broadcasting Network)	Religious and entertainment programming, including morning show (*US AM*), soap opera (*Another Life*), and college football.	4/77
CBS Cable	Music, dance, drama, and special series presented in three-hour blocks and repeated as such.	10/81 (discontinued, 10/82)
CNN (Cable News Network)	News programming featuring newscasts and special-interest series.	6/80
CNN-2 (Cable News Network)	Headline news service to augment CNN-I; provides time for local-news inserts.	1/82
C-SPAN (Cable Satellite Public Affairs Network)	Live coverage of the U.S. House of Representatives, plus special features.	1/79
Cable Text	Transmission of *UPI, Reuters News View, View Weather, Dow Jones,* and *Quotrader.*	10/80
Christian Media Network	Religious programming, films, and family entertainment. CMN also offers music programming.	6/81
Daytime	Programming geared toward women and general-interest features.	3/82
Dow Jones Cable News	Alphanumeric news in fifteen-minute cycles, featuring stock-market and financial news as well as *Wall Street Journal* and *Barrons* articles. Dow Jones also has plans for a two-way service on Westar IV.	4/81
ESPN (Entertainment & Sports Programming Network)	Round-the-clock sports coverage and sports-related programming.	9/79
EWTN (Eternal Word Television Network)	Catholic religious and entertainment programming.	8/81
Episcopal Television Network	Religious programming.	6/81
Financial News Network	News service primarily for broadcast television featuring stocks, financial, and general economic information.	11/81

Table 1–1 continued

Service	*Programming*	*Start-up Date*
Good Stuff	Programming geared toward children between the ages of 7 and 14.	12/81
Lifestyle	Audio service of easy-listening music.	2/81
MSN (Modern Satellite Network)	Daytime public-service and informational programming, including *The Home Shopping Show.*	1/79
MTV (Music Television)	Taped concerts and music programming in Dolby FM.	8/81
NCN (National Christian Network)	Multidenominational religious programming.	6/80
National Jewish Television	Programming for the Jewish community.	5/81
Nickelodeon	Programming for children and adolescents.	4/79
North American Newstime	Slow-scan news service, including TV program guides and *The Women's Channel.*	5/81
PTL (People That Love)	Christian entertainment, news, and specials. PTL also provides an audio service to 500,000 subs.	4/79
Private Screenings	Adult R-rated entertainment.	12/80
Reuters Monitor Service	Encoded-text-retrieval service.	80
SIN TV Network	Spanish-language series, specials, and news. SIN provides international satellite transmission as *Univision.*	9/76
SPN (Satellite Program Network)	Life-style programming, featuring international talk, how-to, business and financial, health programs, and classic movies.	1/79
Satellite Music Network	Popular-music audio programming in stereo.	9/81
TBN (Trinity Broadcasting Network)	Religious programming.	5/73
UPI Cablenews Wire	Alphanumeric specialized news-text service and Spanish-language text.	79
USA Cable Network	General-interest, sports, and entertainment programming, including *Calliope* for children, *The English Channel,* and health and fitness information.	9/80
The Weather Channel	Continuous coverage of national, regional, and local weather.	5/82
WFMT	Chicago's fine-arts radio in stereo.	3/79
WGN	Chicago's independent station, offering movies, sports, news, and specials.	11/78

Table 1-1 continued

Service	*Programming*	*Start-up Date*
WOR	New York's independent station, offering movies, news, specials, and *Madison Square Garden Network* sports.	4/79
WTBS	Turner Broadcasting's Atlanta superstation, featuring movies, sports, syndicated programs, and news.	12/76
Cable Health Network	Viacom programming related to health, nutrition, fitness, and better living.	6/82
Cable Newspaper Corporation	News and special-feature programming drawn from local newspapers throughout the country that have leased-access resources.	9/82
Chromatic Animation	An all-animation service, featuring short subjects, computer graphics, and innovations in animation.	Fall/82
Cinemerica	Drama, variety, specials, education, and sports for subscribers aged 45 and over.	Summer/82
Delta	An all-sports network planned by ABC in conjunction with ESPN.	Fall/82
Heartbeat Media Network	Music/life-style programming covering the full spectrum of music and related areas.	Fall/82
KidVid Network	Daytime programming for children.	10/82
The Nashville Network	Country-music entertainment.	83
National Jazz Network	An audio-only service of jazz music.	3/82
Satellite News Channels	All-news service jointly planned by ABC and Group W.	6/82
Spanish Universal Network	International films, news, and entertainment from six countries.	7/82
UTV Cable Network	*Involvision* programming, encouraging audience involvement—games, home shopping, how-to, and public affairs planned.	Summer/82
Video Concert Hall	An all-music service.	Fall/82

Source: Adapted from National Cable Television Association, "Compensation Plans by Satellite Programmer" (September 1982).

have been available for leased use under a similar arrangement. The programming for leased cable channels can be produced by the leasee. For example, an automotive firm may produce and distribute a weekly program about sports cars. Independent producers have leased time to distribute programming that they finance with advertising revenues. Some industry observers have argued that more local programming should be arranged through public or leased-access channels, leaving the cable operator with the hardware business of running the system. This would encourage a more competitive market for programming.

Pay Cable

Pay cable refers to the distribution by cable of pay-television programming. Although pay cable accounts for about 85 percent of the total pay-television market, competing technologies such as broadcast television or microwave can also be used to distribute pay-television programming to subscribers.

After more than ten years of operation, HBO controls about 60 percent of the pay-television market and contributes about $85 million in annual profit to the parent company, Time, Inc. The market response to these profits has been a rapid increase in the number of pay-television networks. A list of pay-television services appears in table 1-2.

The growth rate of revenues from pay-cable has outstripped the growth rate of basic cable, even though pay cable has been an option available only to basic-cable subscribers. The higher growth rate has been possible because many systems have begun to offer multiple pay-cable services.

Movies represent the most popular type of programming offered by pay cable. The programming mix for HBO is 60–70 percent recent movies, 10–25 percent sports, and about 15 percent special entertainment. So far the pay-television networks have a relatively limited production budget. Most of the movies shown on pay cable are purchased from the motion-picture studios where they were produced for first-run distribution in theaters. The increasing market penetration of pay television is expected to result in larger expenditures on original programming for pay television.

A few of the pay-television networks offer X-rated movies. These so-called blue channels are not subject to the FCC regulations that govern the program content of licensed broadcast stations. FCC regulations do, however, prohibit a cable operator from using access channels to distribute obscene or indecent materials. This restriction applies only to programming subject to the exclusive control of the cable operator. Of course, all such material is subject to legislation regarding obscenity.

Two-Way Services

The cable services based on a two-way flow of information between the subscriber and cable system may be grouped into two categories: video and nonvideo. Video services—which include video games, business information, and consumer services—require a subscriber-friendly terminal—that is, a video terminal and keyboard allowing subscribers to converse with the

Table 1–2
Satellite-Distributed Pay-Cable Services, 1982

Service	*Programming*	*Start-up Date*
AETN (American Educational Television Network)	Geared toward continuing education for licensed professionals.	1/81
Bravo	Performing-arts programming of international film, dance, and music.	12/80
CSN (Concert Satellite Network)	Live coverage of recording acts in concert on a pay-per-view basis.	1/82
Cinemax	Home Box Office's second tier of pay service; all-movie programming.	8/80
Eros	Adult programming featuring R-rated feature films.	4/82
Escapade	Adult (R-rated) films and specials. Escapade carries *Playboy* entertainment programs.	12/80
GalaVision	Movies, variety, and sports specials for Spanish-speaking subscribers.	10/79
HBO (Home Box Office)	Cable's first pay service, offering movies, sports, and specials.	12/75
HTN (Home Theatre Network)	Family entertainment, G- and PG-rated movies. Service ceases on 30 June (see HTN Plus).	9/78
HTN Plus	An expanded family-film and travel service to replace the current HTN after July.	3/82
The Movie Channel	All-movie programming.	4/79
Showtime	Viacom and Teleprompter pay service offering movies, dramatic features, and specials.	3/78
Spotlight	Pay programming for Times Mirror, Cox, TCI, Storer, and Cablevision cable systems.	5/81
Black Music Network	A contemporary music format featuring live concerts, interviews, and video jocks; geared toward an urban audience.	7/82
Buena Vista Channel	Bilingual programming aimed at Hispanic subscribers.	Summer/82
The Disney Channel	Family-entertainment programming drawing on the resources of the Disney Studio.	1/83
Don King Sports & Entertainment Network	Sports, concerts, and special presentations on a pay-per-view basis initially.	Summer/82
The Entertainment Channel	Full entertainment programming featuring films, Broadway theater, family shows, and BBC programs.	6/82
Home Music Store	Multichannel music service that allows subscriber to choose music in catalog fashion.	Summer/82

Table 1-2 continued

Service	*Programming*	*Start-up Date*
Penthouse Entertainment Television Network	Adult entertainment, films, and features.	Fall/82
Video Text	Comprehensive videotext service to include over 5,000 "pages" of information, including each system's local newspaper and catalogued information from Time, Inc.	Fall/82

Source: Adapted from National Cable Television Association, "Compensation Plans by Satellite Programmer" (September 1982).

system. Nonvideo services—which include home-security, medical-alert, and energy-management systems—do not require a subscriber-friendly video terminal. In many cases, nonvideo service may be directly wired into the home in the same way as an electric circuit.

Among the video services, Playcable offers the awesome combination of cable and video games. Subscribers can toy with the maneuvers of simulated star wars by drawing from the many game programs available over the Playcable system.

A variety of interactive video programs have been tested on the Warner-Amex Qube system in Columbus. For example, subscribers have voted to determine how the plot of a melodrama should twist; they have also been electronically polled for their opinions on public issues. Subscribers can use the system to shop for anything from groceries to money-market funds.

Another application of the cable-computer interface is the home information system. An example is The Source, owned by Reader's Digest, which gives cable subscribers access to information stored in a computer. A number of business information systems have also developed. Data Resources, Inc., part of McGraw-Hill, provides access to economic information, computer software, and technical expertise for subscribers. Gnostic Concepts offers computerized access to information relevant to the electronics industry.

In 1980 Satellite Business Systems established a joint-venture partnership with IBM, Comsat, and Aetna Insurance Company to offer data communication and video teleconferencing using cable-television systems for local distribution. Some analysts expect the video telephone to be the major impetus for growth in the cable industry through the 1990s. Banks or corporations with multiple locations, or companies that exchange information with other firms, are potential customers of both data and video networks. Several New York banks are already using the Manhattan cable system for interbank communications.

At RCA an electronic mail system is under development. This expanding variety of services devoted to the electronic delivery of graphic and textual information is called *videotext*. As with many technologies in their formative years, several incompatible electronic designs are competing for use in the videotext market. National standards have not been developed, and a few technological problems remain unsolved. In fact, the United States has been relatively slow in developing videotext. England, Canada, and France have each installed major systems.

Among the nonvideo services offered by cable are burglar alarms, smoke and heat detectors, and medical-alert devices. Often the cable system links these electrical devices with a central computer. The devices are monitored by the computer as it electronically sweeps through the cable system every few seconds.

Public utilities have discovered several energy-saving applications of nonvideo cable services. The cable system can be linked to a household's heating or cooling system, water heater, or other appliances, shutting this equipment down for short periods during peak energy demand. Water and electrical meters can be remotely monitored by the system, sparing utilities the expense of periodic visits to read meters.

When one considers the market for these two-way cable services, it should be remembered that many can be provided by using conventional telephone lines and computers. Of course, cable has the advantage of video capability. Innovative combinations of broadcast signals and telephone lines, however, have the potential to provide many services that are currently the exclusive domain of cable television.

Competing Technologies

Telephone

The recently divested local subsidiaries of American Telephone and Telegraph (AT&T), the local telephone companies, already provide information retrieval and the link to household security systems. Plans are underway for development of energy-management, meter-reading, and videotext systems. As the companies replace copper telephone lines with higher-capacity fiber optics, they could also offer video capability. At that point any cable service could be provided by a telephone company. The extent to which cable and telephone companies battle in the marketplace will be determined largely by regulatory and legislative decisions. Current regulations prevent local telephone companies from providing video services. The telephone companies, with their large assets, long experience, and almost universal market penetration, could pose a significant competitive threat to cable.

SMATV

Satellite Master Antenna Television (SMATV) represents a scaled-down version of a cable system. An SMATV system is an earth station designed to receive satellite-distributed signals that is shared by the occupants of a building or garden apartment. An SMATV operator negotiates an individual contract to wire each building. The SMATV system typically provides fewer services than a regular cable system—usually programming from the superstations and a few pay-television networks. For the operator, SMATV has the advantage of low start-up costs and few regulatory obligations. For the subscribers, SMATV has offered access to a limited number of cable services where conventional cable systems have not been installed.

Recently SMATV operators have been behaving more like private cable companies. A Dallas operator, Video International Productions, plans to offer a sixty-channel SMATV system. The operator will link a number of buildings together by microwave with a central earth station to form an extensive SMATV system. Given the high cost of cable in urban centers, SMATV constitutes a cost-effective alternative to conventional cable.

Several companies have been organized to install earth stations for individual households. Given the high purchase price, however, the market has been limited almost exclusively to areas not served by conventional cable systems. Also, several legal issues remain unsettled concerning the private reception of satellite-distributed signals. Few of the housholds with earth stations have paid for the signals they receive, and it is sometimes difficult for satellite networks to collect from these scattered private users. So far little attempt has been made at collection enforcement. The small size of the market has made litigation and enforcement uneconomical. As the market has increased in size, however, litigation has intensified.

Direct Broadcast Satellite (DBS)

More individuals would buy their own earth stations if the price were lower. The crucial technological-cost relationship for a satellite communications system is that the more powerful (and more costly) the transmission from a satellite, the smaller (and less costly) the earth station needs to be.

The transmitting power of a satellite may be increased in three ways: by increasing the size of the transmitter, by increasing the bandwidth of the channel transmitted, or by focusing the energy of the transmitter on a narrow geographical area. Increasing the size of a satellite transmitter is very expensive, since communications satellites are launched into space at a cost of thousands of dollars per ounce. Increasing the channel bandwidth reduces the number of channels that can be transmitted. Focusing a beam on a small geographic area from the distance of geosynchronous orbit, 22,300 miles, presents several technological obstacles. Nevertheless, the Communications Satellite Corporation (COMSAT) has relied on each of these three means of improving transmitting power, but principally on the focused beam, in developing a system that allows the use of a three-foot parabolic dish as an earth station. For a cost of about $40 per month, subscribers will receive two to six channels from a direct broadcast satellite (DBS). In the near future, however, the relatively high cost of DBS should limit the market to areas in which cable television is not available.

The COMSAT DBS system involves two orbiting satellites, at an estimated cost of $30 million each, for every geographic area served. One of the two is activated, sending signals to earth-station receivers. The other orbiting satellite acts as a backup and is activated if the first one fails.

Orbiting satellites have an excellent record of component reliability. Usually the life of a satellite is limited by the amount of propellant on board, used to align the satellite with the earth. As the earth rotates, it wobbles. In order to maintain the same relative position with the earth, the satellite ejects some propellant, adjusting its orbit. When the satellite runs out of propellant, it begins to drift out of position. Occasionally the technological problems are more dramatic. The communications satellite SATCOM III made headlines in 1979 when it disappeared as RCA engineers in New Jersey assumed tracking responsibilities from the National Aeronautics and Space Administration (NASA). SATCOM III is believed to be orbiting the sun. Several of its transmitting channels had been dedicated to bringing new services to the cable-television industry.

In 1982, the FCC approved plans by several major corporations to develop DBS systems. One plan, by CBS, involves high-resolution broadcasting that improves television picture quality by increasing the number of lines that compose the video images. The system, used in Europe, is not compatible with the television sets owned by most U.S. households. If the system becomes popular, therefore, it could lead to a fundamental change in U.S. television technology.

Teletext

The broadcast version of videotext is teletext. Broadcast stations can include data on the vertical blanking interval of the broadcast signal that is not used to transmit images. Subscribers must have the equipment needed to separate the teletext information from the rest of the broadcast signal. The disadvantages are that teletext allows for only one-way transmission of data and at a considerably slower speed than videotext. Though adequate for the distribution of news or advertisements, teletext could not accommodate information retrieval. Nevertheless, as a result of their relatively low cost, teletext systems have been operated in England for a number of years with considerably more success than the more expensive interactive information-retrieval services. Even so, teletext has not been profitable there.

In the United States, CBS has been most active in developing teletext systems. At KNXT, Los Angeles, CBS broadcasts news, weather, and sports information with teletext. One drawback has been the cost of providing receiving equipment for teletext subscribers. The cost is high, about $200 for each subscriber, in part because teletext equipment must currently be provided separately from the television set. Built-in equipment or attachable components would be cheaper. As is also the case for videotext, one hindrance to providing built-in receiving equipment is the lack of industry standards. Of course, it is risky to establish standards while it remains unclear which technology should be developed.

Multipoint Distribution Service (MDS)

For several decades microwaves have been used to transmit data to business users. An MDS microwave system uses a fixed, omnidirectional transmitter operating in the 2,150- to 2,160- or 2,162-MHz portion of the electromagnetic spectrum. The receivers are usually horn-shaped or parabolic microwave antennas. Thus MDS operates as a closed-circuit microwave transmission system. Little use was made of MDS until 1970, when the allocated bandwidth was widened to allow for transmission of color-television signals. Each color signal requires 6 MHz of bandwidth, so the MDS-allocated spectrum can distribute at most two color-television signals in a given area. The usual range of an MDS system is twenty-five to thirty miles.

MDS has been used by SMATV operators to link several buildings in a network with an individual earth station. Hotels represent a large part of the market for this application. Cable operators sometimes use MDS to link individual cable systems or to link receiving antennas with a cable headend.

About one third of the more than 100 operating MDS systems are used to distribute pay television directly to households. Many of these systems are owned by cable companies. One drawback to MDS is that a microwave receiver must be connected to the set of each MDS pay-television subscriber, resulting in high installation costs. Also, because microwave receivers are available from many electronics stores, MDS operators often broadcast a signal that has been scrambled by an encoding device to discourage households from intercepting the signals without paying the MDS operator. Each MDS subscriber must, therefore, be provided with a device to decode the signal and produce an intelligible television picture.

A technical drawback to using MDS is that microwaves require a line-of-sight transmission path. Buildings and trees deflect the microwave signal. As a result, many of the households in a given area cannot receive MDS programming. A regulatory constraint is that MDS operates as a common carrier, subject to FCC restrictions on use and program content. At present, the FCC limits MDS systems to using less than half of their operating time for broadcasts to affiliated subscribers. For example, an MDS system distributing pay television to household subscribers is required to use or lease more than half its time for some other purpose, such as data transmission among business users. As a result of the technical and regulatory restrictions, MDS account for less than 5 percent of the pay-television market.

A recent CBS proposal to the FCC was aimed at making MDS more directly competitive with cable. The plan calls for reallocating to MDS a little-used part of the microwave spectrum currently assigned to educational institutions. As a use of this bandwidth, CBS proposed to build an eight-channel system in New York and a four-channel system in Philadelphia that would compete directly with planned or existing cable systems. FCC action on the CBS plan and other existing MDS regulations will determine, to some extent, the level of competition between MDS and the cable industry.

Subscription Television (STV)

As distinguished from microwave-based broadcast systems, STV relies on a standard UHF or VHF television signal to distribute programming in a local market. The STV operator, who usually receives programming from a pay-television network via satellite, broadcasts a local television signal that has been scrambled by an encoding device. The advantage of STV over MDS is that a special microwave antenna is not required for reception; all television sets in the local market receive the scrambled STV signal. Also, STV signals are less subject to interference from local terrain.

The STV operator provides each pay-television subscriber with the decoding device needed to interpret the scrambled STV signal. In order to avoid paying the STV (or MDS) subscription fees, a few unscrupulous individuals have acquired a decoding device from someone other than the operator. Use of the devices to pirate pay-television signals is illegal. Catching pay-television pirates is difficult, however, so operators have attempted to outwit them with increasingly sophisticated signal-scrambling technology.

Among the most sophisticated decoder-billing systems is SelecTV, developed by Teleglobe. Subscribers activate the SelecTV decoder to receive specific programs offered by the local STV station. A microprocessor stores a record of the programs that were decoded. Each month this record is transmitted to a central computer via a telephone line. The central computer then bills the subscriber based on the number and type of programs decoded—pay-per-view billing. If the subscriber is delinquent in paying bills, the central computer can send a signal through the phone line to the microprocessor, which deactivates the subscriber's decoder. These provisions against pirating and freeloading have resulted in an interactive STV system combining broadcast and telephone technology. Similar innovations may make STV more nearly competitive with cable.

Historically, STV has not done well when introduced where cable is available. When it has been installed where no cable is available or prior to cable, market penetration has been 10 percent (Los Angeles) or higher (Austin, Texas). Most analysts expect, however, that STV will not pose a significant threat to cable in the near future. It currently accounts for less than 10 percent of the revenues derived from distributing pay television.

The STV industry is still relatively new. Although the FCC began experimenting with a system during the 1960s in Hartford, Connecticut, the first fully licensed station did not begin broadcasting until 1977. About thirty stations were licensed and operating by 1982. As a licensee, the STV operator is subject to FCC rules and regulations, which include ownership restrictions and, when the station broadcasts an unscrambled television signal, programming responsibilities. Several STV stations operate during part of the day as conventional, advertiser-supported television stations broadcasting unscrambled signals. As of 1982, however, STV stations are not required to provide conventional, unscrambled programming, nor are they limited to areas having at least four conventional stations, as they had been previously.

Like any broadcast station, STV operates on an assigned frequency. Of course, most of the good television frequencies have already been assigned. In major markets, therefore, STV has been limited to UHF frequencies. Sometimes the use of UHF results in poor signal reception. To resolve this problem, television translators have been used to extend the effective market area of STV stations. The recent FCC action to license thousands of new low-power television stations will serve to increase the potential number of STV stations.

Low-Power Television (LPTV)

Broadcasting with less power than a conventional television station, LPTV stations serve a smaller market area. Depending on local terrain, an LPTV station may serve an area within a fifty-mile radius, although some will be licensed to broadcast over only a six-mile radius. More conventional stations broadcast over an area with a radius of at least seventy miles. The advantages of an LPTV station include lower start-up costs and fewer regulatory responsibilities than for a conventional broadcast station. Also, the geographical packing of LPTV stations into urban markets will allow for broadcasting on previously unassigned local UHF and VHF channels. These advantages have inspired a flood of more than 6,000 applications for new LPTV licenses since 1980.

The first of the new LPTV stations went on the air in December 1981. UHF channel 26, operating in Bemidji, Minnesota, broadcasts both STV and conventional programming to a market area about forty-five miles in radius. Start-up costs amounted to about $450,000, compared with the approximately $2 million start-up costs for a full-power station. An LPTV station could be started for as little as $20,000.

Before 1981 there were already 4,000 LPTV stations operating as television translators around the country. They were used as relay stations to extend the broadcast areas of a conventional television station. Extending this concept and taking advantage of the lack of ownership restrictions, Sears-Roebuck, through its Allstate Insurance Company subsidiary, plans to use almost 100 UHF translator stations to form a network over which it will broadcast KUSK-TV of Prescott, Arizona. This translator network, to be called the Neighborhood TV Company, will distribute rodeos, country music, and cowboy humor to most parts of the country.

Although much of the FCC's motivation in awarding the new LPTV licenses was to encourage the local origination of programming, many of the applicants propose to construct STV stations. Whether developed primarily as advertiser-supported or as pay-television stations, LPTV may provide competition for local cable operators.

The Economic Interdependence of Cable and Broadcast Television

Selling time for commercial advertisements provides revenue for a conventional broadcast television station. A direct relationship exists between revenue and the size of the audience a station reaches. A 10-percent reduction in audience results in approximately a ten-percent reduction in revenue (Charles River Associates 1978b). To the extent that cable systems divert audience from local broadcast stations, advertising revenues of broadcast stations are diminished. To the extent that they increase audience size, revenues are increased.

To determine the extent of this interdependence, the FCC conducted a series of economic inquiries during the 1970s. The inquiries were intended to assess the potential effect on local broadcast stations of eliminating rules, then enforced, prohibiting the use of microwaves to import distant broadcast stations into television markets. Several econometric studies were undertaken. In general, it was determined that cable would increase the revenues of stations whose signals were imported into other markets because audience size for these stations would be increased. Rolla Park of the Rand Corporation forecast a 3-percent gain for urban UHF independents and a maximum 11-percent loss for stations experiencing new competition from imported signals (Park 1971a). According to a study by Wharton Econometric Forecasting Associates, the potential audience loss for some local stations could be quite large (Schink and Thanawala 1978). For example, in a smaller market with three VHF and one UHF independent, the UHF station could suffer a 40-percent loss of audience if the cable system imported six distant signals.

A Brookings study concluded that the only losers would be ABC, CBS and NBC, and their affiliates. Further, these losses were essentially the benefits of monopoly power reflected in the high profits of network-affiliated stations (Noll, Peck, and McGowan 1973). For example, of 244 VHF network-affiliated stations operating in the top 100 television markets during 1976, income before federal tax for the year represented 125.1 percent of the original cost of tangible property less depreciation. For VHF independents the figure was 25.5 percent. Fifty-two independent UHF stations operating in the top 100 television markets during 1976 generated $10.3 million in losses. According to the Brookings study, many of these UHF independent stations should welcome cable. Local cable distribution would improve reception of the notoriously poor UHF signals and so encourage larger audiences.

Cable, then, operates with a complementary-competitive relationship to broadcast television. As the cable industry relies increasingly on non-broadcast sources for entertainment services—pay television, satellite-distributed networks—the competitive aspect of the relationship between cable and broadcast television will intensify.

Home Video Equipment

Both cable and broadcast television face potential competition from home video equipment. Although home video dates back to the 1950s, video technology has been substantially improved, and several large companies are devoted to expanding the market. Two distinct types of technology are represented, videotape and videodisc. The technology of videotape is essentially the same as that of audio recording. Electric current is used to align molecules of a magnetic coating on a strip of plastic tape. The market offers about five different and incompatible types of videotape technologies. A video player machine may be used only to play prerecorded tapes; it cannot record. The video recorder can both play and record tape. Often, the household use of a video recorder has been to store programs distributed by cable or broadcast television. A series of court actions has begun to test the legality of this use.

The videodisc player is a more recent technology than tape, dating from 1974 in the U.S. market. The concept of the videodisc is essentially the same as that of the phonograph record. Information is imprinted onto a plastic disc using a molding process. As with videotape, there are approximately five different and incompatible technologies for videodisc. One of the basic differences among them is the use of either a physical-contact system (similar in function to the phonograph needle) or the use of a laser to read the disc. The physical-contact systems are generally cheaper. The laser, on the other hand, produces no disc wear and allows random access, an advantage if the disc is to be used for information storage or retrieval.

Compared with videotape, the major disadvantage of videodisc is that it cannot be used to record. Prerecorded discs are distributed through retail stores and through direct mail order. Disc-of-the-month clubs have already been organized. Books, newspapers, magazines, and the videodisc rely on the physical distribution of imprinted information. These media represent the third player in a technological competition among broadcast, cable, and physical-distribution systems. Economics and regulation will determine the market for each system.

Summary

Cable television originally served as a community antenna. Four categories of equipment included in the system design are: the headend, the distribution plant, the subscriber interface, and the studio. Two categories of service are basic cable, or regular service, and pay-cable, which offers programming from a pay-television network. Broadcast television has historically been the communications technology most closely linked to cable. Competing technologies include telephone, Satellite Master Antenna Television (SMATV), Direct Broadcast Satellite (DBS), Teletext, Multipoint Distribution Service (MDS), Subscription Television (STV), Low Power Television (LPTV), and home video equipment.

2 Regulatory Development of the Industry Structure

Cable television combines two traditionally distinct telecommunications-industry structures: mass communication and intercommunication. Mass communication, particularly broadcast television and radio, has provided a one-way distribution of programming to a large audience. Intercommunication, particularly telephone and telegraph, has provided a two-way distribution of information among those having access to the intercommunications system.

The distinction between mass communication and intercommunication is enforced by the Communications Act of 1934, the legislative document on which the structure of the telecommunications industry was predicated. Under Title II of the act, many intercommunications services have been regulated as common carriers, subject to Federal Communications Commission (FCC) jurisdiction over rates and services. According to the act, a communications common carrier is a firm whose services are open to public hire for handling interstate or foreign communications by electronic means. Intrastate communications are generally under the jurisdiction of state or local government.

In recent years the scope of common-carrier regulation has been diminished. The courts and the FCC have acted to remove restrictions on entry into interstate markets and to reduce rate regulation of interstate carriers.

Title III of the Communications Act of 1934 provides for regulation of mass communication through control of the radio spectrum. Broadcast television and radio stations are not deemed to be common carriers; rates have not been regulated. In order to structure the broadcast industry and ensure that programming is in the public interest, the FCC has relied on its authority to license the use of the radio spectrum.

Cable television has been regulated, to some extent, as both an intercommunications (Title II) and mass-communications (Title III) industry. Cable systems provide services similar to those offered by both these industries. The leased-access channels and nonentertainment two-way services, such as data retrieval or video telephone, fit the intercommunications model. The retransmission of broadcast signals and the distribution of pay-television more closely fit the mass-communications industry model. A brief history of the development of these services will reveal the interplay of economic and regulatory forces that have structured the cable-television industry.

Telephone

Among the intercommunications industries, the market for telephone service has been by far the largest. The first telephone patents issued in the United States in 1876 established Bell Telephone as an industry monopolist. Bell's patents prohibited competitive entry by other firms. As a result, and as intended by the patent laws, the company earned a rate of return well above the national average. For taking the risks associated with introducing a new technology, investors were rewarded with a 46-percent rate of return from dividends alone.

The patents expired in 1894. Soon afterwards, more than 6,000 independent firms entered the market. One effect of the resulting competition was that some areas were served by two or even three independent companies. Individuals wanting full service—the capability of connecting to every telephone in the area—had to subscribe to more than one company. Having to pay more than one monthly telephone bill has obvious disadvantages.

Response to the effects of competition came from both public and private quarters. In 1907 the financial interests of J.P. Morgan took over Bell Telephone, now American Telephone and Telegraph (AT&T), installing Theodore Vail as president. Under Vail's direction, AT&T began to buy out competing companies and to reestablish the telephone monopoly. At the same time, several state governments reacted to the problem of overlapping systems by requiring their interconnection and by granting a single, franchised monopoly in each geographic area.

In 1910 Congress passed the Mann-Elkins amendment to the Interstate Commerce Act, granting regulatory authority over interstate communications to the Interstate Commerce Commission. Executive branch pressure for system interconnection resulted in the 1913 Kingsury Commitment, named for the AT&T vice-president who negotiated the agreement with the attorney-general's office. Reflecting the antitrust reform of the Wilson administration, the Kingsury Commitment required AT&T to interconnect with all independent companies. It also required AT&T to divest itself of holdings in Western Union, the telegraph company, and to refrain from further acquisition of independent telephone companies. Since that time, mergers and consolidations have reduced the number of independent telephone companies.

Market forces and regulatory decisions acted to reestablish telephone-market monopolists in particular geographical areas. Bell Telephone's operation as a monopoly prior to 1894 demonstrated that, left to itself, a monopolist can make large profits by charging high rates for telephone service. At one time the high profits could be justified as a reward for technological innovation, a justification that expired with the patents. In order to keep rates reasonable, the state governments that had created franchised telephone monopolies began to regulate telephone rates.

The philosophic method of the rate-making authority exercised during this period was formalized in the 1930 Supreme Court case of *Smith* v. *Illinois Bell Telephone Company* (283 U.S. 133). The court held that the local telephone network—or *loop* as it is called—is used jointly by both long-distance and local service. The court's decision required that part of the cost of the local loop be recovered from long-distance revenue. The problem of allocating these joint costs has occupied rate makers for many years. An economically similar problem exists in the relationship between the local cable operator and the pay-television network.

Aggressive regulation of the interstate telephone market began with the Communications Act of 1934, which created the Federal Communications Commission (FCC). The act transferred regulatory authority over communications from the Interstate Commerce Commission to the FCC. It imposed the following duties on interstate intercommunication firms available for public hire, which were considered to be communications common carriers.

Common-Carrier Obligation

1. To furnish service on reasonable request.
2. To file tariffs with the Federal Communications Commission (FCC) specifying rates, practices, and regulations.
3. To obtain FCC permission in order to "discontinue, reduce or impair service to a community or part of a community."

The Communications Act of 1934 conferred the following powers on the FCC:

FCC Common Carrier Jurisdiction

1. To void carrier charges, classifications, regulations, and practices it finds unreasonable or discriminatory.
2. To prescribe just, fair, and reasonable charges, regulations, and practices.

From 1934 to the late 1960s the FCC determined reasonable charges for telephone service on an informal basis by monitoring company profits. In

1967 the FCC began to prescribe a maximum allowable rate of return. Disputes arose over what constituted a fair rate of return and on what base the rate of return should be calculated.

Though closely regulated by the FCC, the telephone industry enjoyed virtual freedom from competition in the voice-communication and telephone-equipment markets for more than thirty years. The first major challenge to the protected status of the telephone monopoly came in 1966, when Carter Electronics, a small Texas manufacturing company, brought a private antitrust suit against AT&T and General Telephone and Telegraph (GT&T). The object of the suit, the Carterphone, provided an acoustic interface connecting private radio-communications systems with the telephone network. AT&T declined to interconnect, citing an existing tariff provision against foreign attachments. The FCC responded in 1968 by repealing the provision. A new tariff, which permitted attachments that would not be mechanically harmful to the telephone company, was afterwards adopted. The decision opened the telephone-equipment market to competition.

In 1974 the Justice Department brought suit against AT&T for monopolization of the long-distance telecommunications market. The 1982 settlement of the suit required AT&T to divest itself of the local Bell Operating Companies (BOCs). The local companies continue to be regulated by state commissions but will provide access to the local loop for all long-distance companies. In effect, the long-distance market is undergoing economic deregulation.

Cable Television and Telephone

Coaxial cable, which is commonly used to distribute television signals through a cable-television system, was not developed primarily for that purpose. AT&T has been using coaxial cable for high-capacity telephone links between major cities since 1936. In that year, New York and Philadelphia were connected in the first practical test of coaxial-cable technology.

Restricted from operating cable-television systems by an FCC common-carrier consent decree, AT&T began stringing coaxial cables and leasing them to operators. Many small operators in the capital-intensive cable industry welcomed the AT&T leasing scheme, which reduced their initial capital outlay. Another advantage for cable operators in the AT&T leasing scheme was reduced regulatory restrictions. Local regulation of cable systems is accomplished through the municipal jurisdiction over the use of public property. Cable operators, who leased directly from AT&T, could avoid dealing with the local regulatory authority.

The rift between the leasor and the leasees developed when cable operators began to realize the potential of new markets, such as interactive nonvideo service. These markets were AT&T's traditional domain. AT&T leasing contracts prohibited them from expanding into these new markets. If operators attempted to circumvent the leasing contract by buying their own cable, AT&T would often deny them access to the poles on which cable could be strung. Erecting new poles could be prohibitively expensive or infeasible. In an unregulated market, the firm that owns the telephone or utility poles can potentially extract a very high rent from cable operators. To protect cable operators from the possible monopolistic pricing of pole attachments, the FCC may use its authority to intervene in negotiations when prices become unreasonable. The states may, at their discretion, preempt the FCC's regulatory authority over pole attachments. In fact, the FCC only reluctantly assumed regulatory responsibility. A recent staff report commented:

> The nature of pole attachment rates clearly indicates that regulation should be vested with those persons most familiar with the local environment within which public utilities and cable television systems operate. Given that the determination of reasonable rates, in many instances, will depend on the expected cost of increasing available pole space at some point in the future, familiarity with a utility's local operating environment consititutes a vital input into the decision making process [U.S. FCC 1977].

The potential of new markets for telecommunications services that originally inspired the disputes between AT&T and the cable operators was not lost to the independent telephone companies. Unrestricted by the FCC common-carrier consent decree that had prevented AT&T from entering the market, the independents had begun building and operating cable systems. In 1968 United Utilities and Continental Telephone filed tariffs with the FCC that proposed that they provide a single coaxial cable to carry all communications and information servcies, such as telephone, picturephone, data-transmission, and protection services. The response of the FCC has been to discourage telephone companies from constructing cable systems, except in rural areas.

As part of a 1982 antitrust settlement, AT&T has been allowed to enter the cable-television market. The divested Bell Operating Companies (BOCs), however, are not. The extent to which telephone companies can compete with cable operators continues to be a source of judicial and legislative controversy.

Microwave

The FCC has control over the use of microwaves through its authority to license the radio spectrum as granted by the Communications Act of 1934.

> It is the purpose of this Act, among other things, to maintain control of the United States over all channels of interstate and foreign radio transmission; and to provide for the use of such channels, but not the ownership thereof, by persons for limited periods of time, under licenses granted by Federal authority, and not such license shall be constructed to create any right, beyond the terms, conditions, and periods of the license [47 U.S.C., sections 201–204].

Microwaves occupy the radio spectrum above 890 MHz. Because they are very short waves and accordingly have a high capacity to carry information, they can be used to carry anything from telegraph to television. As a result, microwaves have been used to provide intercommunication and, more recently with MDS, mass-communications services. As might be expected, telephone and telegraph common carriers opposed the private development of microwave, which represents a competing technology. One argument raised was the potential for so-called cream skimming, the notion that the privately controlled new technology would make services available to only the most profitable components of the existing market. The less profitable business would then be left for the common carriers, who are legally obliged to serve the public.

In 1959 the FCC examined the microwave industry on the basis of spectrum utilization and economic impact on competing common-carrier services. The FCC decided in favor of developing private microwave but declined to establish a specific guideline for the interface of microwave and common-carrier service. This issue was to be decided on a case-by-case basis.

One of the longest and most expensive legal battles in U.S. history ensued when Microwave Communications Incorporated (MCI), which provided microwave services to common carriers, petitioned the FCC to be interconnected to the AT&T local telephone-distribution system. MCI planned to offer long-distance telephone service in direct competition with AT&T. In 1980 MCI was granted the interconnection request. Since then MCI has grown dramatically, and other competing long-distance firms have entered the market. In order to resolve the cream-skimming issue, AT&T has slowly been relieved of common-carrier economic obligations, allowing it to compete head-on with the new long-distance companies.

The use of satellites to relay microwave communications over long distances was proposed as early as 1945 by British science-fiction writer Arthur C. Clarke. On 10 July 1962 the United States launched the first communications satellite, Telstar, which had been developed by Bell Laboratories. Shortly thereafter, Congress created the Communications Satellite Corporation (COMSAT). The corporation, responsible for the creation of an international satellite communications system, is owned half by private firms and half by the general public. The FCC regulates COMSAT under both the Federal Communications Act and the Satellite Communications Act.

In 1972 the Supreme Court ruled that any financially and technically qualified entity could use satellites to provide specialized microwave services. Shortly thereafter, RCA entered the market with the SATCOM series of communications satellites. The use of satellite-distributed signals has since been a key factor in the growth of cable. Direct broadcast satellite (DBS) services promise to make COMSAT a competitor in the mass-communications market.

Mass Communication

In 1899 the Marconi Company of America, a subsidiary of the British Marconi Wireless Telegraph Company, became the first company to provide radio service in the United States. The U.S. Navy, which at the time had some experience with the new technology, attempted to organize a company and to prevent the sale of certain key U.S.-owned patents to the Marconi company. The Navy was unsuccessful in organizing either a government-controlled monopoly or a private corporation. Instead, a consortium composed of AT&T, General Electric, and the United Fruit Company combined financial resources and patent rights to organize the Radio Corporation of America, RCA.

The structure of the broadcast market as envisioned by the consortium resembles more the noncommercial public broadcast system than the commercial market of today. Community leaders were to form broadcasting associations, which would provide programming and pay for use of broadcast facilities operated by AT&T.

A 1926 agreement among the founders of RCA divided the industry into now familiar segments. RCA was to operate the broadcasting stations but would exclusively use AT&T lines to connect stations in the broadcast network. In a sense, AT&T has since operated a national cable-television system by connecting the broadcast network.

A crucial series of decisions in the structuring of the broadcast industry were made in the late 1940s for radio and the early 1950s for television. It was then that the FCC decided how the available radio spectrum would be allocated for commercial use. Two plans were considered. One, proposed by the now defunct Dumont Television Network, envisioned a small number of very powerful stations that would provide service to virtually everyone in the country. The FCC chose an alternative, the 1952 Allocation Plan, whereby a large number of small stations would provide local television.

One drawback to the 1952 Allocation Plan, anticipated by the FCC, was that some rural communities, lacking the population base to support a local station, would not receive a television signal. The drawback proved to be more severe than expected. Fewer than half the 2,000 stations envisioned by the FCC actually went on the air. In urban markets, however, more stations were proposed than channel space had been allocated.

Television stations, licensed for operation by the FCC under Title II of the Communications Act of 1934, must demonstrate on applying for a renewal of their operating license that they have been responsive to community needs. STV stations broadcasting a scrambled signal and LPTV stations have been exempted from this requirement.

Both radio and television were allowed to develop as advertiser-supported media, free of common-carrier obligations and rate regulations. The FCC contended that enough stations would develop in each locale to encourage competitive rates. In general, however, network-affiliated stations in the top 100 markets have been very profitable; rates, therefore, have been above competitive levels.

In order to promote a diversity of program content, the FCC has issued licenses, renewable at least every three years, with strict requirements concerning cross-ownership of other communications media and multiple ownership of radio and television broadcast stations. Empirical support for the deterioration of information diversity as a result of ownership concentration may be found in a study by W.T. Gormley (1976). An increase in the cross-ownership of local television stations and daily newspapers in a local market was found to reduce the number of news stories covered, as well as to reduce the frequency and diversity of editorial opinion. In a 1973 study, B.M. Owen found that the price of advertising space was about 10 percent higher for newspapers cross-owned with a local television station than for those not cross-owned (Owen 1973).

The development of cable systems by companies already owning a local media business is forbidden by existing regulations. The number of cable systems a company may own in different geographic markets is not limited. A company owning more than one cable system is commonly referred to as an MSO, multiple system operator. The number of radio and television stations a company may operate is limited, but the FCC has considered proposals to ease or abandon the ownership restrictions. Few companies other than the three major networks (ABC, CBS, and NBC), actually own the maximum five VHF television stations allowed by the FCC regulation. The networks, with their combined fifteen owned and operated stations (O&Os), account for approximately one-third of the industry's pretax operating revenues. Of the remaining stations, about 85 percent are affiliated with one of the networks.

Although the three networks employ their own lobbyist in Washington, the formal primary lobby of the industry is the National Association of Broadcasters (NAB). The NAB has actively lobbied for the strict regulation of cable television. In counterpoint, the National Cable Television Association (NCTA) has lobbied for the deregulation of cable.

Federal Regulation of Cable Television

Entrepreneurs like John Watson were free to develop cable television without federal regulation until 1962. Before that time, regulation was a state or local option. Many systems were franchised by municipalities, with subscriber fees fixed in terms of the franchise. Some states passed legislation treating cable systems as common carriers to be regulated by public-utility commissions. Other systems developed free of any government intervention.

Federal regulatory intervention into the cable industry was prompted when cable operators began importing television broadcast signals from outside the local market. One issue that was raised involved the copyright laws allowing cable operators to import signals without paying for the imported programming. The local broadcasters argued that they suffered an unfair competitive disadvantage, since a significant portion of their operating expenses were devoted to programming costs.

The owners of some broadcast stations whose signals were distributed to cable systems outside their market also complained. Although broadcast stations received no direct compensation for the use of their programming, program suppliers, who had entered into contracts with these broadcasters for distribution in a specific geographic area, wanted compensation for the additional geographic distribution of their programs by cable. The program suppliers, unable to collect from the cable operators, tried to collect from the broadcast stations. Local advertisers, on the other hand, were reluctant to pay increased rates to reach distant audiences unlikely to buy their products. The broadcast stations were caught in a financial squeeze.

The FCC reacted in 1962 by prohibiting a cable operator from using microwaves to import a distant television signal into a community already supporting a local television station. In its decision against the Carter Mountain Transmission Corporation (32 FCC 479), the FCC asserted both its jurisdiction over microwave systems and its authority to base actions on the economic interdependence of several communications technologies. The courts upheld the FCC's decision (321 F 2d, 1962; 375 U.S. 951, 1963).

In the mid-1960s the National Association of Broadcasters (NAB) and the National Cable Television Association (NCTA) began a series of unsuccessful negotiations to resolve the copyright issue. In 1966, anticipating a negotiated settlement, the FCC established regulations for all cable systems (14895 FCC 2d 725, 1966). It was required that cable systems distribute the signal of all television stations operating in the local market. According to the 1966 regulations, distant television signals could not be imported by a cable operator into a top-100 market without an FCC hearing to determine if the action would serve the public interest. This regulation placed the burden of proof on the cable operator. Further, the requirement of a hearing was not limited to signals imported by microwave; all imported television signals, regardless of the transmission medium, were subject to FCC approval.

With the 1966 regulations, the FCC had again significantly increased the range of its jurisdiction. The copyright dispute had provoked FCC action, but the underlying motivation was a desire to maintain the communications-industry structure that was designed into the 1952 Allocation Plan. The objective of the plan was to encourage local broadcast service. Cable systems, by importing distant television signals, seemed to promise the national distribution of individual television stations. National television stations had been considered, but rejected in favor of localism, when the 1956 Allocation Plan was conceived.

The FCC's authority to impose the regulations on cable was upheld in the 1968 Supreme Court case of *The United States* v. *Southwestern Cable* (392 U.S. 157, 1968). The decision conferred on the FCC jurisdiction over cable to the extent that it was "reasonably ancillary to the effective performance of the Commission's various responsibilities for the regulation of broadcasting." Later that year the courts upheld the FCC's authority to enforce restrictions on signals carried by cable systems in the case of *Black Hills Video Corp.* v. *FCC* (399 2d 2d 65, 1968). In the same year, a Supreme Court decision complicated the copyright issue. In *Fortnightly Corp.* v. *United Artists Corp.* (392 U.S. 390, 400, 1968), the Court held that the retransmission of broadcast signals by cable television did not constitute a "public performance for profit" under copyright law. That is, cable operators were not required to compensate broadcast stations whose programming the cable system distributed.

The 1966 FCC regulations, which required a case-by-case hearing to authorize importation of distant signals, quickly proved too cumbersome. The FCC was flooded with hearing requests. Years would have been required to make a determination on each one. Since the profitability of major-market cable systems depended significantly on the distribution of imported broadcast signals, many cable operators delayed construction while waiting for an FCC hearing determination.

Acting from the broad authority conferred on it by the *Southwestern* cable case, and having decided that a comprehensive set of rules was required to govern the hearing process, the FCC ordered a total freeze on all construction in the top 100 markets. The freeze gave the FCC time to organize.

The 1972 Cable Television Rules (36 FCC 2nd 143)

After almost fours years of study, the FCC issued its most comprehensive set of rules governing cable and its relationship to broadcast television. The requirement of an individual federal hearing for each franchise award was replaced with a certification process. Before beginning operations or adding a new broadcast signal to existing systems, cable operators were required to file a Certificate of Compliance indicating that their cable systems complied with FCC rules.

The 1972 FCC rules established standards of technical performance and reliability for cable systems. The rules also more closely defined the regulatory relationships among federal, state, and local governments. Basic-cable rates were to be under the jurisdiction of state and local governments. Pay television would be the responsibility of the FCC. The division of regulatory authority reflected the local, or intrastate, character of cable systems and the national, or interstate, character of pay television. Historically, the federal government has been responsible for interstate communications. The FCC did, however, limit municipal franchise fees to 3 percent of cable-system gross revenues. Previously, fees as high as 40 percent of gross annual revenue had been required by municipalities.

The 1972 rules contained much detail regarding the importation of distant television signals. In the top 50 markets, cable operators were limited to importing signals that would, in addition to the local television market, provide a total of three network affiliates and three independents. In the next-largest 50 markets, the combined total of broadcast and cable was to be three network affiliates and two independents per market. In all other markets, three network affiliates and one independent were to be allowed, unless the quota was filled by the local broadcast market. In that event, two imported signals were allowed to cable systems in the top 50 markets; one would be allowed in the next 50 markets. An additional provision required that a signal imported by cable from the top 25 markets be from the closest city, and that if three signals were imported, one must be UHF.

Two rules enforced the exclusive rights to programming broadcast by local television stations. A cable system could not import the programming of a network station that was being carried simultaneously by a local broadcast affiliate of the same network. Cable systems in the top 50 markets were also forbidden to telecast a syndicated show within a year of its release in a syndicated market or where a local station had bought the rights to the program.

Minimum requirements were established by the FCC for the local franchising of cable systems. New systems were required to have at least twenty channels of capacity, one nonbroadcast signal for every broadcast signal carried, and two-way capability. Systems with more than 3,500 subscribers were required to have at least one access channel filled with a significant amount of local programming. Cable operators were to provide the equipment needed to produce locally originated programming. In the top 100 markets, four access channels were required: one for the public, one for educational purposes, one for government use, and one for leased access.

Court Challenge to the 1972 Rules

While the FCC was taking a protectionist attitude toward broadcasting, the courts generally acted to encourage competition. In the same year, 1968, that the *Southwestern* case acknowledged the FCC's authority to regulate cable, the Department of Justice Task Force on Communications Policy submitted a report to President Johnson advocating that cable television be allowed to develop freely as a communications medium in competition with the established broadcasting industry (LeDuc 1973).

In 1972 Midwest Video, a large multisystem cable company, challenged the FCC regulation requiring cable systems serving more than 3,500 subscribers to originate programming. The lower court's decision in favor of Midwest Video was overturned by the Supreme Court in a five-to-four decision (*U.S.* v. *Midwest Video;* 406 U.S. 649, 1972). Writing the opinion of the Court, Justice Brennan held that the earlier case of *Southwestern* had established FCC jurisdiction over cable.

Midwest Video, however, had not seen its last day in court. In what has come to be called Midwest II (*FCC* v. *Midwest Video Corp.;* 440 U.S. 689, 1979), the cable operator successfully appealed to the U.S. Court of Customs and Patent Appeals regarding the FCC's mandatory access rules. In his decision for the court, Judge Markey asserted that the FCC's access and mandatory channel-capacity regulations exceeded the commission's jurisdiction. The judge acknowledged the earlier Supreme Court decision (Midwest I) but drew an important distinction between the two cases. In the earlier case, he observed, the FCC had required cable systems "to act like broadcasters and originate programming." In Midwest II, however, the mandatory-access provision required cable operators to "act like common carriers." A common carrier must be available for public hire. The effect of Midwest II, overturning the common-carrier obligations, was to rescind the 1972 access and channel-capacity regulations.

The authority asserted by the FCC to regulate all signals distributed by cable was modified in the 1976 decision of *NARUC* v. *FCC* (533 F 2d 601, 1976). It declared that the FCC could not exercise jurisdiction over intrastate two-way, point-to-point nonvideo cable transmissions. These data and other nonvideo transmissions, however, are subject to state regulation.

A Federal Circuit Court opinion (*Home Box Office* v. *FCC;* 434 U.S. 829; 1977) rescinded the FCC regulations governing pay-cable content. The court did not revoke FCC jurisdiction but contended that the FCC had failed to demonstrate that a problem existed that required regulation. The philosophy that regulation should be curtailed unless clearly required has since been adopted at the FCC. It has resulted in reduced regulatory intervention. For example, the FCC does not regulate pay-cable rates, although it has maintained the authority to do so. The FCC's authority to regulate pay-cable rates was affirmed in the 1978 Circuit Court case of *Brookhaven Cable TV* v. *Kelly* (573 F 2d, 1979). The court decided that the FCC's jurisdiction over pay-cable rates preempted rate regulation by the New York State Commission on Cable Television.

Executive-Branch Activity

In 1971 the Office of Telecommunications Policy (OTP) inaugurated a study of appropriate regulation for cable television with the formation of the Cabinet Committee on Cable Television. By 1974 basic legislation was proposed by the OTP. The legislation was simultaneously attacked by the FCC as being too inflexible for economic purposes and by the Justice Department as being too flexible for economic purposes (Hallowell 1980). One issue raised by the OTP's accompanying 1974 Report to the President of the Cabinet Committee on Cable Communications has yet to be completely resolved:

> Control of cable facilities should be separated from control of programming and other services provided over the channels on those distribution facilities. . . . By separating the distributional function in cable, which is a natural monopoly, from the programming functions, which can be highly competitive, the dangers of intrusion and influence in programming can be avoided while the wide variety of competitors vying for the public's attention can be expected to produce a diversity of programming [Hallowell 1980].

By 1975 the proposed legislation of the office was revamped, and the decision became one of either adopting the comprehensive legislation or allowing the industry to be deregulated. In support of the deregulatory forces was a 1976 memorandum to the Domestic Council Review Group on Regulatory Reform in which Paul MacAvoy of Yale noted that "the introduction of new technology invariably erodes the position of established firms, but this effect generally benefits the public in terms of new products and competitive prices" (MacAvoy 1977). Though noting the problems inherent to deregulation, MacAvoy's memorandum expressed the growing deregulatory mood in Washington.

Congressional Activity

In his concurring opinion to the Midwest I case, which first upheld the FCC's jurisdiction over cable, Chief Justice Burger declared that

> the Commission's position strains the outer limits of even the open-ended and pervasive jurisdiction that has evolved by the decisions of the Commission and the courts. The almost explosive development of CATV suggests the need for a comprehensive re-examination of the statutory scheme as it is related to this new development, so that basic policies are considered by Congress and not left entirely to the Commission and the courts. [*US* v. *Midwest Video,* 406, US 649, 1972]

Congress had considered the problem. Since the FCC began regulating cable in 1966, and until the mid-1970s, Congress had tried and failed three times to pass comprehensive cable legislation. In 1976 a staff report of the House Subcommittee on Communications, *Cable Television: Promise vs. Regulatory Performance,* proposed to determine what legislative action should be taken. With the release of the report, the subcommittee, chaired by Lionel Van Deerlin, embarked on an extensive revision of the Communications Act of 1934.

The bill, the Communications Act of 1978 (H.R. 13015), as proposed by Congressman Van Deerlin, precluded any federal regulation of cable television. The FCC would have been replaced by the dramatically weaker Communications Regulatory Commission, and the public-interest standard would have been abolished. The bill was not passed into law. A rewrite of the Communications Act of 1934 appeared again the following year as H.R. 6121, the Communications Act of 1979, but again failed to pass into law. Since that time, Congressman Van Deerlin, a major force in the rewrite effort, has not returned to the House.

The Senate Communications Subcommittee had held hearings on cable television beginning in 1977. Communications legislation was introduced into the Senate in 1979 (S. 611 and S. 622) and in 1980 (S. 2827). Both the Senate and the House introduced similar deregulatory communications legislation in 1981 (S. 898 and H.R. 5158). The approved provisions regarding cable included pole-attachment regulation and restrictions on the ownership of a rural cable system by the dominant local telephone company. In 1982 Senator Barry Goldwater introduced comprehensive cable-television legislation as the Cable Communications Act of 1982 (S. 2172). The bill would restrict cable companies from leasing channels for nonentertainment services, a move to restrict the intercommunications services similar to those provided by common carriers.

Although through 1982 Congress had been unable to pass comprehensive cable legislation, two important amendments were added to existing laws. In 1976 the Copyright Law was amended (Public Law 94-533, Title 17) to require payment of a copyright fee by cable operators for the right to retransmit radio and television signals. Congress also added amendments to the Communications Act of 1934 (P.L. 95-234, 1978). Subject to preemption by state regulation, the FCC is required to regulate the rates, terms, and conditions for pole attachments.

The Federal Regulatory Environment to 1983

By 1975 the FCC had begun to modify the cable-television regulations devised in 1972. In September 1977 the FCC eliminated all its franchising standards—required access programming and minimum channel capacity—except for the limit on municipal fees. The comprehensive certification procedures that had been in force were replaced in 1978 by a few simple registration requirements. In fact, the FCC no longer examines municipally negotiated franchise agreements for compliance with federal rules.

During the mid-1970s the FCC conducted an economic inquiry into the relationship between cable and broadcast television. Finding little evidence of significant economic reasons to restrict the importation of distant signals, the FCC removed most of the restrictions on their use by cable operators. As of 1978, broadcasters have the burden of proof to demonstrate that the importation of distant signals would be harmful. The rule change became effective in 1981 when the courts affirmed the FCC's decision in the case of *Malrite T. V. of New York* v. *FCC* (652 F 2d 1140, 1981). By 1983 only the following federal regulations remained in force.

Local Signals. The cable operator must carry, on request, all licensed television stations within thirty-five miles of the community served by the cable system and any other stations significantly viewed in the community. Television translator stations of 100 watts or more that serve the community must also be carried on request, but STV stations broadcasting a scrambled signal need not be. The rule applies to all Grade B noncommercial educational stations and Grade B signals of all stations serving communities in small television markets (a Grade B signal is one that can be viewed by at least 50 percent of the television households in a market). Legislation has been proposed to eliminate the local-signal carriage requirement.

Network-Program Nonduplication. A cable system may not import network programming that is being carried simultaneously by a local broadcast station affiliated with the network.

Retransmission. Cable operators may not tamper with the broadcast signals they distribute. For example, a cable operator may not replace a commercial advertisement of a retransmitted signal with one of its own, thereby collecting advertising revenue.

Technical Standards. Cable operators are required to conduct annual performance tests to ensure system safety and reliability.

Pole Attachments. Rates charged by local utilities and telephone companies for use of poles to support cable-television distribution plants are subject to FCC regulation unless regulation is preempted by the state. According to the Communications Act of 1934, as amended, rates should be set at reasonable levels.

Copyright. As of 1978, cable operators are required by copyright law to pay royalties for the privilege of retransmitting television programs. The fee is determined by the gross revenue of an individual cable system and the number of distant broadcast signals it distributes. Copyright fees are likely to remain controversial through the 1980s. Although broadcasters have won a significant battle in the legislation that enforces the copyright requirement, the broadcast industry maintains that it is still subsidizing cable. Broadcasters note that whereas programming costs average one-third of yearly expenses for a broadcast station, programming costs are an insignificant percentage of expenses for most cable systems. By comparison, the total copyright payments of the cable to the broadcast industry during 1979 were less than the programming expenses of one independent New York television station. Cable operators argue, however, that by retransmitting broadcast signals they have increased the audience—and hence the potential advertising revenues—of broadcast stations.

Ownership. A telephone company may not own a cable system in the area in which it also operates the local telephone exchange. A broadcast station may not own a cable system in its broadcast market. Television networks may not own cable systems. All these restrictions are, of course, subject to an FCC waiver granted on a case-by-case basis. For example, CBS, prohibited by the rules from owning a cable system, obtained a waiver to operate one in California. No restriction exists on the number of cable systems a single firm may own.

Equal Opportunity. Cable operators are required to offer equal employment opportunities for all individuals. Systems employing five or more people must file a copy of their equal-employment-opportunity plan with the FCC.

Municipal Franchise Fee. A maximum of 3 percent of gross revenues may be charged as a fee by municipal authorities granting franchise agreements. A waiver will be granted by the FCC to allow fees in the 3- to 5-percent range if the higher fee is needed to cover the local cost of regulation. The FCC has recently studied several proposals for eliminating the franchise-fee limitation. At present the fee limitation represents the most significant FCC intervention into the local regulation of cable. Although the FCC has the authority to regulate pay-cable rates as determined in the *Brookhaven* case, it does not currently exercise that authority. The regulatory development of cable television is, therefore, almost exclusively the domain of state and local governments.

State and Local Regulation

Municipalities were the first to impose regulations on cable television as a consequence of jurisdiction over the use of streets and rights of way. In the 1968 case of *TV Pix, Inc.* v. *Taylor* (396 U.S. 556, 1968) the U.S. Supreme Court upheld the constitutionality of state regulation of cable television. The court emphasized the local character of the industry.

At least forty states enforce one or more cable laws, which commonly deal with franchising procedures, theft of service, pole attachment, taxation, or rate regulation. By 1979 some form of rate regulation had been imposed by at least ten states: Alaska, California, Connecticut, Delaware, Hawaii, Massachusetts, Nevada, New Jersey, New York, and Vermont. Often the state public-utility commission is charged with the responsibility of regulating rates.

The most popular form of rate regulation is rate of return, used in Alaska, Connecticut, Hawaii, Nevada, and Vermont. A 20-percent rate of return is allowed in Vermont. Hawaii uses a rate-of-return criterion that varies in accordance with the circumstances of local systems as determined by the regulatory commission. In Delaware, any price increase exceeding 5 percent must be approved by the state regulatory commission. New Jersey employs a common-tariff approach whereby maximum price increases are specified for several types of systems that provide community-service channels. The state commissions in Alaska, Massachusetts, and New York all deregulated rates or related procedures in decisions issued in 1980.

In all but a few states, cable operators must enter into negotiations with municipal authorities over the terms of franchised use of municipal streets and rights of way. As a result, the municipal franchise agreements represent the most pervasive control over the development of cable television. Municipalities have even successfully circumvented the intent of the federally regulated limit on franchise fees. Lump-sum cash payments have been negotiated by some cities. For example, San Antonio, Texas, recently received $1 million for the franchise award. Other cities have received free use of expensive telecommunications equipment. For many financially strapped municipalities, franchise deals represent an economic blessing. A 1978 Harvard study, however, advises against special cable-television taxes that create a disadvantage for cable companies not shared by other businesses (Levine, Kalba, and Hochberg 1978).

The municipal franchise agreement represents a contract negotiated between the city and a cable company. In order to accomplish effective negotiations, both sides must have definable goals and pertinent information. The following two chapters include information about the market for cable services. Using this information, later chapters examine the practicability of achieving private and municipal goals through the franchising process.

Summary

Many of the federal regulations on cable television were attempts to define the economic interrelationships among cable and other communications technologies, especially broadcast television. Since the 1970s the courts and the FCC have rescinded most of the regulations regarding cable-television programming and franchising. Among the most important remaining federal regulations are the required copyright payments for the use of broadcast signals by cable operators and the limitation on cable-television franchise fees.

Public responsibility for the development of cable-television systems rests primarily with municipalities and other local jurisdictions. The most common form of local control is the franchise contract, a regulatory descendant of the franchised telephone monopolies that developed early in the century.

3 The Cost of Cable Television

During the early 1980s annual operating expenses for a cable system averaged between about $70 and $150 per subscriber, depending on system size and type of programming offered. Capital expenditures averaged between $300 and $1,000 per subscriber, depending on the channel capacity of the system and the type of construction required. As a consequence of the high initial capital costs, four or five years may be required before a new cable system begins to generate a positive cash flow. High capital costs may also prevent competitive entry once a system is established.

One of the most widely debated issues concerning the economic character of the industry has to do with the similarity between a cable system and a public utility. Is a cable system a natural monopoly? If so, regulation may be required to induce cable operators to provide service at reasonable rates. For example, the local telephone exchange, thought to be a natural monopoly, has been regulated in most states since the early 1900s to ensure that telephone service is provided at reasonable rates.

The cable industry has argued against the public-utility model, preferring to characterize itself as an electronic newspaper. The economic implication is that cable operators, like newspaper owners, should not be rate-regulated. Other industry arguments against regulation have included the nonhomogeneous nature of cable services, the notion that cable services are nonessential goods, and the availability of close substitutes such as subscription television (STV). These precepts, which focus on the demand side of the industry and are discussed in the next chapter, have found many adherents among cable operators and at the FCC.

Those favoring the public-utility model include the National League of Cities, which characterized the cable industry as an electronic pipeline and a de facto monopoly. The league reports that for only 8 of the 5,000 systems operating in 1981 did one cable operator compete with another for the same subscriber (National League of Cities 1981). In about twenty cities that have had two competing systems, one system has gone out of business, leaving the other as a monopoly.

From the perspective of costs, a natural monopoly is said to exist if the average cost of production declines as output increases. Such a production process exhibits declining average total costs, which means that costs are

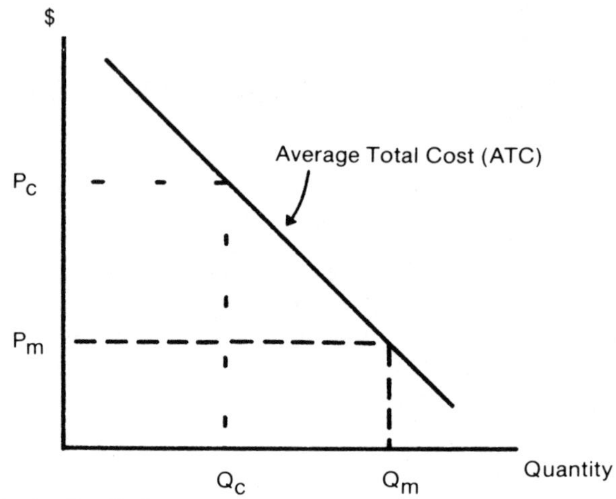

Figure 3–1. Declining Average Total Cost for a Natural Monopoly

lower if service is provided by one firm rather than by any combination of firms. The firm characterized by figure 3–1 exhibits declining average total costs. That is, the per-unit cost of producing output Q_m by a monopoly firm is lower than the cost of producing output Q_m by having two firms operate at Q_c, where $Q_c + Q_c = Q_m$. Including normal profits as part of costs, P_m would be the price charged by a monopolist earning normal profits. The higher price P_c would be charged by each of two firms operating at Q_c. The difference between P_m and P_c is the per-unit saving that can be achieved by allowing monopoly production.

In order to determine whether cable television is a natural monopoly, a detailed engineering and econometric study of the cost of cable television was conducted. The results, reported in this chapter, indicate that cable systems do exhibit declining average total costs for the number of channels distributed, the number of subscribers, and the size of the geographic area served.

Equipment Requirements for a Cable-Television System

A cable system receives, amplifies, and distributes signals to subscribers on the system. The equipment used in a cable system is divided into four general categories.

1. The *headend,* which receives and processes signals.
2. The *distribution plant,* commonly coaxial cable, which carries signals to and from locations in the area served by the system.

3. The *subscriber interface,* which connects subscribers to the distribution plant.
4. *Cablecasting equipment* used for local program origination.

Representative costs for this equipment are contained in table 3–1. The general structure of a cable system is illustrated in figure 3–2.

Table 3–1
Cable-Television Equipment Costs
(1982 dollars)

Headend Equipment Requirements and Costs[a]		
A. Required for each cable system		
Tower	$25,000	
Building	10,000	
Standby power	5,000	
Program switcher/splitter	30,000	
Interconnect receiver/transmitter (for multiple-franchise interconnect)	3,000	
Pilot carrier generator	5,000	
Over-the-air antenna (for broadcast market) and amplifier	10,000	
FM receivers	5,000	
B. Required for each channel		
Signal process/modulators (with pass-band filter)	$1,500	per channel
Spare signal process/modulators (with pass-band filter)	1,500	for each four channels
Signal scramblers	1,000	per pay-cable channel
C. Microwave		
Earth station (for satellite-distributed signals; one per satellite)	$20,000	for each station (plus $0.10 per subscriber per month fee to satellite owner)
Receivers	4,500	for each channel
Antenna (land-based microwave)	15,000	
Radio equipment (for land-based microwave)	9,000	per channel per hop
Standby power	4,000	
Distribution Plant Costs (Includes installation)		
40-channel capacity aerial cable (36 downstream, 4 upstream)	$10,800	per mile
40-channel underground cable (36 downstream, 4 upstream)	20,000	per mile
Pole rental	160	per mile annual expense
Shadow trunk	1,800	per mile
Franchise interconnect (for multiple franchises)	5,500	per mile

Table 3–1 continued

Subscriber Interface Costs	
Drop (including installation, splitter, and bridger amplifier)	$45.00 per subscriber
Converter	
(nonaddressable)	65.00 per subscriber
(addressable)	125.00 per subscriber
Other Equipment Costs[b]	
Preoperating capitalized labor	$500,000
Tools and test equipment	2% of headend equipment; plus $125 per mile
Motor vehicles	12,000 each; one per 5,500 subscribers plus one per 50 miles of cable
Office space	7,500 plus $500 per 1,000 subscribers in excess of 5,000
Furniture and office equipment	7,500 plus $2,500 per 1,000 subscribers

Note: Typical depreciation lives for capital equipment:

15 years—antennas, towers, earth stations, microwave, headend components, distribution plant, and preoperating capitalized labor.

10 years—converters, leasehold improvements, furniture, office equipment, and program-origination equipment.

5 years—tools and test equipment.

3 years—motor vehicles.

[a]Costs based on figures distributed by Teleprompter, Inc.; Merrill Lynch, Pierce, Fenner & Smith, Securities Research Division; and Scientific Atlanta.

[b]Costs are estimated from the New Jersey sample in the form suggested by Comanor and Mitchel (1971) and indexed to 1982 dollars using the relevant producer price index.

The Headend

A schematic of the headend appears in figure 3–3. Two outside signal sources represented in the figure as being introduced into the headend are off-the-air broadcast and microwave. Local broadcast signals of sufficient strength to ensure reliable signal quality may be received by an off-the-air television antenna. VHF and UHF television antennas can simultaneously receive several signals, but they are highly directional and sensitive to signal strength. The required number of off-the-air antennas for each cable system depends on the number and type of signals to be received and the direction of their origination. The required height and location of the tower on which antennas are mounted depends on conditions of the local terrain and the distance to the signal source.

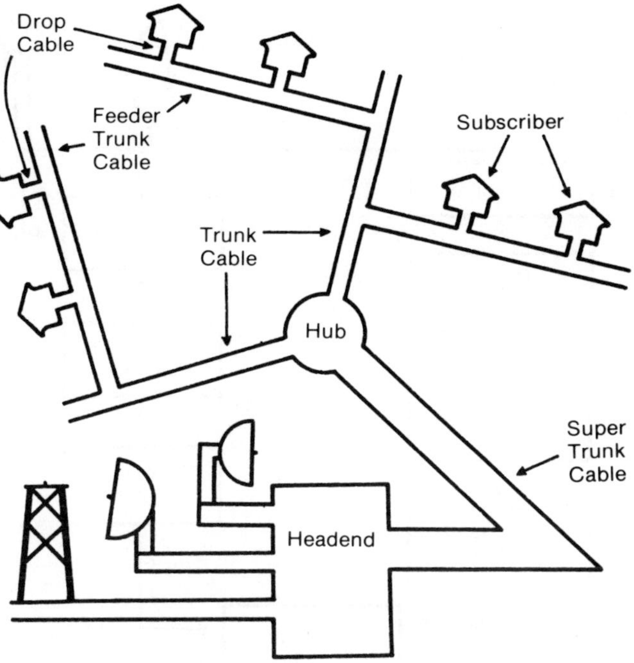

Figure 3–2. The Cable-Television System

Microwave antennas, also highly directional and sensitive to signal-strength variations, are of two general types depending on the signal source: land-based transmitter or communications satellite. Since microwaves travel in straight lines, the curvature of the earth restricts a land-based microwave signal from being received at a distance of more than seventy miles without retransmission. Usually spaced at twenty-five- to thirty-mile intervals, each retransmission, or hop, requires a station that includes a signal receiver/transmitter, or *transponder*. Because retransmission stations are expensive, land-based microwave is most often used for transmitting signals over relatively short distances.

Satellite links are used for longer distances. The downlink receiver—satellite to earth station—includes a parabolic dish antenna, ten to twenty feet in diameter, and associated electronic equipment. Since communications satellites are in geosynchronous orbit, always at the same relative position in the sky, the earth station antenna need not be equipped with expensive tracking equipment to keep the antenna focused on the satellite. When installed, the earth-station antenna is aligned with the satellite.

Several satellites, each with a twenty-four-channel capacity, are currently used to distribute signals to cable operators and for other communi-

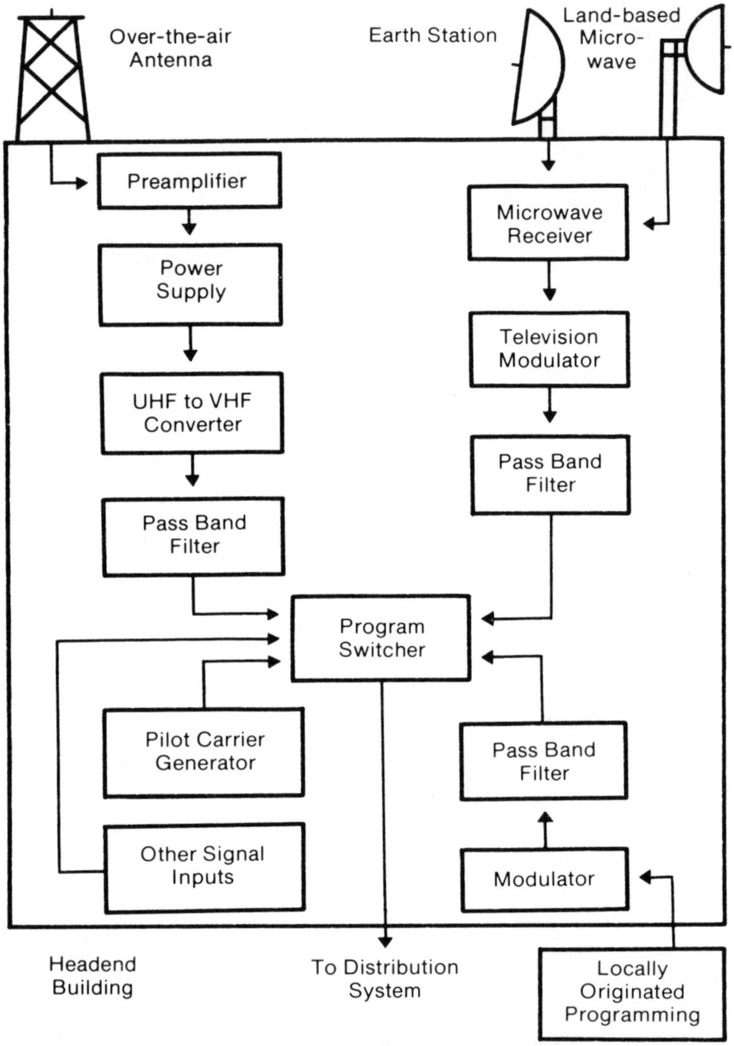

Figure 3–3. The Headend of a Cable-Television System

cations purposes. Satellites can be spaced at about every 4 degrees from 95 to 130 degrees, on the horizon. Closer spacing begins to produce signal interference. A separate parabolic dish antenna is currently required to receive signals from each satellite; however, new designs being developed would allow a single earth station to receive from several satellites.

Received signals must be processed for distribution. A pass-band filter is used to remove signal noise. The pilot carrier generator controls gain. A switching device, the directional coupler and splitter, allows the cable operator to edit and channel signals into the distribution system.

Signals of pay-television services are often scrambled at the headend to prevent unauthorized viewing. A descrambler must then be installed for each pay-television subscriber. Although signal scrambling can be relatively successful in preventing freeloaders from watching without paying, it requires expensive equipment at both the headend and the subscriber interface. A less expensive way to prevent unauthorized viewing involves installing a filter at the subscriber interface which blocks pay-television from entering unauthorized sets. Innovative freeloaders, however, have discovered that the filter can be disconnected with a minimum of technical expertise and a little stealth.

A more elaborate pricing scheme, pay-per-view, can rely on a computer to electronically sweep through the system every few seconds to record the program selections of each subscriber and thus provide accurate information for billing purposes. As cable systems increase their interactive services, additional computer capacity and switching equipment will be required at the headend. For example, video-telephone service, if offered on a cable system, would require switching equipment at least as sophisticated as is currently found in the local loop of a telephone system. Systems that allow subscribers to respond to programs or to make purchases from their homes must be equipped with computing equipment to process subscribers' decisions. In general, as cable systems offer more sophisticated services, the cost of headend electronics will increase.

One aspect of sophisticated cable services that often forms the basis of debate over appropriate communications technology may be summarized by a reference to George Orwell's novel *1984*. In that book an omniscient communications technology was employed by a villainous totalitarian government to keep everyone under surveillance almost all the time.

Cable Distribution Plant

The distribution plant is the major capital expenditure for a cable system. It includes the cable and associated electronic equipment designed to distribute signals throughout the cable-system service area. The cost of the distribution plant depends on the channel capacity, the length of the plant, and the type of construction required for installation—underground or aerial.

Most of the systems in operation were built with coaxial cable, a thick copper wire sheathed in plastic and aluminum to reduce interference and corrosion. The maximum channel capacity of a cable depends largely on the diameter of the copper core: a thicker core allows more capacity. The effective channel capacity of the cable also depends on how successfully the aluminum and plastic sheathing protects against signal interference. Often, a few channels cannot be used because local broadcast signals penetrate the sheathing and create interference. During the late 1970s and early 1980s, the industry standard was 40-channel (300-MHz) coaxial cable. One channel is defined to be the band-width required for a broadcast signal. A thicker, 54-channel (400-MHz) capacity cable has been developed more recently. The total channel capacity of a system can be increased beyond 54-channel limit by increasing the number of cables in the system; for example, two 54-channel cables would be used to construct a 108-channel system.

The channel capacity of a single cable may be divided between downstream (from the headend to the subscriber) and upstream (subscriber to headend) channels. Two-way coaxial cable requires diplexing filters to separate upstream from downstream channels; consequently, it is more expensive than one-way cable.

Two-way cable also necessitates separate amplification for upstream and downstream signals, further increasing the cost. Amplification is required because electronic signals traveling through cable attenuate rapidly. The acceptable distance between amplifiers decreases as channel capacity increases. An amplifier is required every 2,000 feet for a 20-channel system and every 1,200 feet for a 40-channel capacity system.

The sequential spacing of amplifiers along a coaxial cable is referred to as a *cascade*. Because signal quality deteriorates with repeated amplification, the maximum number of amplifiers in a cable-system cascade is limited to about twenty-five or thirty. Further amplification would reduce signal quality below technically acceptable standards. Consequently, the diameter of a 40-channel cable system is limited to between five and seven miles. Beyond that range, another headend is required. Improvements in amplifier technology will increase the maximum cascade length.

Like the spokes of a wheel, the cable of the distribution plant extends outward from a central hub. The hub is usually near the center of the franchise area in order to reduce the required amplification throughout the system. Headend equipment designed to receive signals, however, need not be located at the hub of a cable system if a different location offers better reception, as is often the case. For example, antennas can be mounted on nearby mountains to reduce interference. If so, the hub and headend are connected by coaxial cable, the so-called supertrunk, or sometimes by land-based microwave. A splitter divides the supertrunk into as many trunk cables as extend outward from the hub. To reach all of the neighborhoods in the service area, trunk cable may be split into feeder trunk.

Included in the cost of cable is the capitalized labor associated with distribution-plant construction. Aerial construction involves stringing the cable from utility poles or building attachments. Cable operators rent these poles from the local utility at rates that are subject to state or federal approval. Often rental payments to building owners are also required.

More expensive is the underground construction required where aerial attachments are unavailable or illegal. Since aerial cables are unattractive, the proportion of aerial to underground construction sometimes becomes an issue in awarding cable franchises. During the early 1980s underground construction amounted to approximately one-quarter of the total miles of cable installed. The cost of underground construction varies considerably depending on local terrain. In San Francisco underground-construction costs have been reported to be as high as $100,000 per mile. Where the underground conduits of local utilities can be used to accommodate cable, the cost of underground construction may be moderated.

When it is anticipated that system channel capacity will be expanded in the future, the long-term cost of underground can be reduced by installing a *shadow trunk*—a cable without electronic components such as amplifiers or filters—at the same time that the regular distribution plant is installed. The shadow trunk can be activated as needed by supplying the electronic components, thus avoiding the cost of a major excavation subsequent to the initial installation. For ordinary aerial construction, however, no appreciable cost savings are realized by such anticipatory construction. (Malarkey, Taylor, and Associates 1975).

One disadvantage of anticipatory construction is that the installed technology may be obsolete by the time it is called into service. For the technologically dynamic cable industry, selecting the appropriate level of anticipatory construction presents a difficult, if not impossible, problem. For example, fiber optics will eventually replace coaxial cable as the best technology for the distribution plant. Fiber-optic cables are made from glass fibers and rely on light to transmit signals, rather than on the electric current used in coaxial technology. As a result, signals delivered through optic fibers are not affected by electrical storms. They also require less amplification, which reduces costs and results in less signal distortion with distance. In Baltimore County, Maryland, thirty-two miles of coaxial supertrunk were recently replaced with fiber, reducing the required number of amplifiers from forty-seven to thirteen. Fiber-optic cables have been used to connect telephone centrals four or five miles apart without intermediate amplification.

Unlike coaxial cable, fiber optics do not corrode and so will extend the useful life of cable systems. Perhaps the most remarkable advantage of fiber optics, however, is channel capacity. A fiber-optic cable of the same diameter as a coaxial cable used to distribute 40 channels will provide 1,000 channels of capacity.

Fiber optics have not yet replaced coaxial cable in local cable systems because light signals must be transformed into electrical signals to allow for both amplification and switching in fiber-optic cables. This requirement increases technological complexity and cost. Although fiber-optic cables are currently cost-effective for long-haul transmission with infrequent switching, the cost is prohibitive for a cable distribution system wherein signals must be switched at frequent intervals from the feeder-trunk distribution cable into the homes of subscribers. Fiber optics have, however, been used as supertrunk in cable systems in which frequent switching is not required.

The Subscriber Interface

The electronic switching device, or splitter, used to divert signals from the distribution plant is called a *top*. A bridger amplifier boosts the diverted signals into the drop cable that leads into a subscriber's residence. The converter is the final set of components needed to introduce the cable-system signals into a subscriber's television set. Included in the converter is a transformer, which makes the cable signal compatible with the subscriber's television set, and a channel selector. For systems distributing scrambled pay-cable signals, a descrambler may be included in the converter. An addressable converter is installed with systems offering two-way services; it allows subscribers to send signals upstream through the distribution system to the headend.

Programming Costs

Most of the signals distributed on a cable system are provided to the cable operator on a contractual basis or in accordance with copyright laws. A few of the signals may carry locally originated programming and therefore require that the cable operator invest in equipment and incur an annual operating expense. Table 3-2 summarizes the costs of programming available to cable operators: retransmitted television signals, satellite-distributed basic-cable programming, pay-television, and locally originated programming.

Table 3-2
Programming Costs for Cable Television, 1982

A. Average costs for locally originated programming	
Equipment	
Public Access	$50,000 for each access studio
Local programming	$300,000 for each channel
Operating expense (technicians, talent, videotape, and miscellaneous)	
Public access	$30,000 for each access studio
Local programming	$220,000 for each channel
B. Copyright fees for retransmission of broadcast stations	
0.799% of gross revenues for the 1st station	
0.503% of gross revenues for the 2nd, 3rd, and 4th station	
0.237% of gross revenues for all additional stations	

Table 3–2 continued

C. Basic-Cable Service	*Cost to System*	*Local Ad Availabilities*	*Compensation to System*
ACSN (Appalachian Community Service Network)	$.05/5,000 or fewer subs; $.045/5,000–50,000 subs; $.025/ 50,000–100,000 (per month)	None	Tuition-sharing plan: $3.30 compensation for each subscriber enrolled in local college course via ACSN[a]
ARTS (Alpha Repertory Television Service)	Free	None	None
BET (Black Entertainment Television	$.01/sub/mo.	6 min./day	$.005/sub/month made available for advertising coop
CBN (Christian Broadcasting Network)	Free	7 news and sports spots daily; additional spots planned	None
CBS Cable	Free	3–4 min./day	$.15/sub for promotion provided[b]
CNN (Cable News Network)	$.15/sub/mo. $.20/sub/mo. (without WTBS)	48 min./day	$.30/sub for promotion provided (up to $10,000)[c]
CNN II	24 hours free with CNN ($.05 if carried alone)	48 min./day	Additional local ad time is available for system producing own twice-hourly news updates[d]
C-SPAN (Cable Satellite Public Affairs Network)	$.03/sub/mo.	None	None
CABLE Health B Network	Free	2 min./hour	Will pay $.15/sub/. yr. (up to 50,000 subs); $.25/sub 50,000–250,000) subs; and $.50/over 250,000). These represent ''homes to be passed, not actual online subs when the contract is signed''[e]
Daytime	Free	2 :30 spots/hr.	Provides matching promotional funds; up to 10,000 subs; $750 10,000–25,000 subs; $1,500 25,000–50,000 subs; $2,500 50,000+ subs: $3,500

Table 3–2 continued

C. Basic-Cable Service	*Cost to System*	*Local Ad Availabilities*	*Compensation to System*
Dow Jones Cable News	$.01/1 sub—up to 15,000; $.03–.04/ sub—15,000–30,000 $.005/sub— 30,000+	None	None
EPG (Electronic Program Guide)	$125/week + $4 per channel—over 14 channels	40 local ad spots/ week	None
ESPN (Entertainment and Sports Programming Network)	$.04/sub/mo.	80 min./day	For systems with 5-year contracts, with full-time carriage (or, on any tier, if at least 90% of subs get the service). $.10/sub/ year compensation provided.
EWTN (Eternal Word Television Network)	Free	None	None
Financial News Network	Free	49 min./progr. day	Revenue sharing is being considered— up to 10% of FNN's annual gross may be distributed.[f]
Good Stuff	$.005/sub/mo.	None	None
MSN (Modern Satellite Network)	Free	2 min./progr. day	None
MTV (Music Television)	Free	48/min./day	Operator may obtain additional revenue from charges for stereo hookups to service.[g]
NCN (National Christian Network)	Free	None	None
NJT (National Jewish Television)	Free	2 min./hour	None
Nickelodeon	$.15/sub/mon.	None	None
North American Newstime	$.03–.05/sub/mon.; $.07 tiered	144 min./day	None
PTL (People That Love)	Free	None	None
SNC I (Satellite News Channels)	Free	2 min./hour	$.50/sub compensation SNC $1 for SNC I and SNC II[h]

Table 3-2 continued

C. Basic-Cable Service	*Cost to System*	*Local Ad Availabilities*	*Compensation to System*
SIN TV Network (Spanish International Network)	Free	12/min./hour	System compensation quarterly for providing affidavit of Spanish-surnamed subs: 1st quarter–0; 2nd quarter–$.025/sub; 3rd quarter–$.05/sub; 4th quarter–$.10/sub. SIN pays out approx. $15,000/mo[i]
SPN (Satellite Program Network)	Free	48/min./day	When SPN moved to Westar 3 (its previous satellite), its distributor, Satellite Syndicated Systems, gave away or leased earth stations to facilitate and encourage the move for systems.[j]
Spotlight	Varies	None	None
TBN (Trinity Broadcasting Network)	Free	None	Beginning in March 1981, TBN purchased $10,000 earth stations for systems with at least 5,000 subs that would carry service full time (after 5 years, system take over earth-station ownership).
UPI Cable News	Approximately $.04/sub/mo.[k]	Programming may be preempted for local ads at any time.	None
USA Cable Network	$.11/sub/mo. ($.05–$.40/sub if tiered)		None
The Weather Channel	Free	48/min./day	Systems with at least 6,000 subs may receive a STAR (Satellite Transponder Adressable Receive) unit free. Operator may attach local tags to national ads with unit and program commercial crawls and provide local news updates.[l]

Table 3-2 continued

C. Basic Cable Service	*Cost to System*	*Local Ad Availabilities*	*Compensation to System*
WGN-TV	$.10/sub/mo. (discounts are available) + Copyright fee	None	None
WOR-TV	$.10/sub/mo. + Copyright fee	None	None
WTBS-TV	$.10/sub/mo. + Copyright fee	None	$.25/sub/yr. in promotional funds provided
D. Pay Cable Service			
BRAVO	Approximately $2.45–$3.95 per sub/mo.	None	None
CINEMAX	Approximately $3–$4 per sub/mo.	None	None
EROS	Approximately $1–$1.50/sub/mo.	None	None
Escapade	Approximately $2.70–$3.25 per sub/mo.	None	None
HBO (Home Box Office)	Approximately $4–$5/sub/mo.[m]	None	None
HTN Plus (Home Theatre Network)	Approximately $2.95[n]	None	None
The Movie Channel	Approximately $3.50–$4.50[o]	None	None
Showtime	Approximately $4–$5/sub/mo.	None	None

Source: Adapted from National Cable Television Association, "Compensation Plans by Satellite Programmer," Washington, D.C. (September 1982). Also see notes.

[a]*Cable Age,* 16 November 1981, p. 39.

[b]*Cable TV Advertising,* 28 July 1981, p. 2.

[c]*Cable TV Programming,* 31 March 1982, p. 2.

[d]*Cable TV Programming,* 31 March 1982, p. 2.

[e]*Cable TV Programming,* 19 April 1982, p. 5.

[f]*Cable TV Advertising,* 25 November 1981, p. 4.

[g]*Cable TV Advertising,* 28 July 1981, p.2.

[h]*Communications Daily,* 25 March 1982, p. 5.

[i]*Cable Age,* 21 September 1981, p. 22.

[j]*Cable Age,* 25 January 1982, p. 14.

[k]*Home Video Report,* 24 August 1981, p. 4.

[l]*Sat Guide,* May 1982, pp. M27–M28.

[m]*Cable Age,* 14 December 1981, p. 7.

[n]*Sat Guide,* December, 1981.

[o]*Cable Age,* 14 December 1981, p. 7.

Retransmitted television signals still provide most of the programming offered on basic-cable systems. For many years cable operators were not obliged to reimburse the broadcast television industry. As of 1978, each cable operator pays an annual copyright fee.

Satellite-distributed programming, such as Nickelodeon, is provided to cable systems for a few cents per subscriber per month. Several satellite networks, especially evangelical networks, provide programming at no cost to the cable operator. In a few cases the cable operator may profit by selling local ads to fill spots in satellite network programs.

The financial arrangements between broadcast stations and the commercial television networks served as the original model for the arrangement between the local cable operator and pay-television network. The commercial television network takes about 30 percent of the broadcast station's advertising revenue from network-provided programming. Pay-television networks typically take 35–50 percent of the monthly subscription fees received by the local cable operator, but the fee structure for several major cable systems has been changed to a per-subscriber basis. The motivation and implication of the per-subscriber fee structure is discussed in chapter 5.

The cost of locally originated programming can vary dramatically among systems, depending on the type and quality of programming. For example, the equipment in an access studio that is available to the public at little or no charge may consist of an inexpensive black-and-white video camera, some electronics connecting the camera to the cable system, and a room with a few bright lights. A part-time technician may be available to advise citizens on the use of equipment. To make franchise proposals more attractive to city officials, some cable operators have promised expensive video equipment and technical expertise.

An example of inexpensive local production is the use of a black-and-white video camera to record the action of an amateur sports event or theatrical presentation. Mounting programs with the production quality of a local broadcast station's evening news requires an investment in at least several hundred thousand dollars worth of studio equipment, including video cameras, video tape recorders, lights, and a studio set. Annual salaries for studio technicians range from $18,000 to $50,000. Salaries for personalities often range much higher, even for local markets.

Operating Expense

Average annual operating expenses for a hypothetical urban cable system are represented in table 3-3 as estimated by the Federal Communications Commission (U.S. FCC 1977). About two-thirds of the annual operating

Table 3-3
Annual Operating Costs for a Typical Urban Cable System
(thousands of 1978 dollars)

	Costs at Market Penetration of		
Category	*30%*	*40%*	*50%*
Wages and salaries—technical	$270.8	$335.6	$402.8
(% of total)	(38.6)	(40.8)	(42.1)
Wages and salaries—administration	167.8	196.6	235.0
(% of total)	(23.9)	(23.9)	(24.6)
Power	39.0	39.0	39.0
(% of total)	(5.6)	(4.7)	(4.1)
Repair/maintenance	30.0	30.0	30.0
(% of total)	(4.3)	(3.6)	(3.1)
Marketing	94.5	94.5	94.5
(% of total)	(13.5)	(11.5)	(9.9)
Postage/billing/office	84.1	112.1	140.1
(% of total)	(12.0)	(13.6)	(14.7)
Insurance	14.5	14.5	14.5
(% of total)	(2.1)	(1.8)	(1.5)
Total annual operating costs	$700.7	$822.3	$956.0

Note: Figures are based on FCC estimates.

expense of a cable system represents labor, and about two-thirds of that figure represents technical labor. The technical labor force of a cable system normally includes a chief systems engineer and support staff of technicians who maintain the cable distribution system. The required number of line technicians, who maintain the trunk cable, is commonly assumed to be a linear function of the miles of distribution plant in operation. The required number of service technicians, who concentrate on the subscriber interface, is assumed to be linear function of the number of subscribers on the system (U.S. FCC 1977). Another 16 or 17 percent of operating expense includes equipment for technical and administrative personnel—office supplies and tools.

The preparation of a franchise proposal, lawyer's fees, travel, and other miscellaneous expenses required as part of the firm's participation in the local franchising process represent one-time operating expenses. Historically, the figure has been as high as $500,000 for each participating firm. Unexpected delays in the bidding process can be expected to raise the figure.

Little opportunity for substitution of inputs, in the economic sense, is available to the cost-minimizing cable operator. In general, the cable operator chooses an acceptable level of system reliability and staffs accordingly.

Statistical Cost Functions for Cable Television

Unlike the case of many other regulated or previously regulated industries—trucking, for example—detailed financial information for the cable and broadcast industries is not easily acquired. The annual cost data published by the FCC are aggregated by state for several system sizes. These data provide only a general idea of cable-system costs, precluding reliable estimation of costs for a specific system. Even among these aggregated data, declining average total costs are evident. Although the FCC does possess detailed operating-cost data for most systems in the country, the information is considered proprietary and not publicly available. New Jersey, however, makes cost information for individual cable systems publicly available. The state requires that cable operators file an annual financial report with the state public-utility regulatory agency.

Using a 1979 New Jersey cross-section of cable systems, equations were specified that estimate annual operating and depreciation expense. The estimated equations, which are reported in table 3–4, include variables for the size of the operating plant and number of subscribers served by the system. The appendix at the end of this chapter provides a brief discussion of the statistical methods employed in estimating the equations.

One of the more interesting results to be inferred from the equations is that costs do not increase proportionately with system length. The exponent 0.71, on the variable miles in equation 1 for annual operating expense in table 3–4, suggests that there are operating economics of scale for cable systems. Doubling the number of miles in a cable system should less than double total operating expense. As a result, monthly subscription prices for large systems should be lower than for small systems offering equivalent services.

The incremental operating expense of adding a subscriber to the system is, according to the estimate, about $13 per year. The incremental depreciation expense for each subscriber, from equation 2 in table 3–4, is about $16 per year. These values for incremental expense are especially important to the efficient pricing of cable services, the topic of chapter 5.

Also of interest is the fact that operating expense, which does not include programming costs, was not measurably different for a 12- or a 36-channel system. This result indicates that the average operating expense per channel declines as the number of channels increases.

According to equation 3.2, the annual depreciation for headend and distribution-system components would amount to about $536 per mile for a

Table 3–4
Estimated Cable-System Costs
(1982 dollars)

Equation 1: Estimated cable-system annual operating expenses
(excluding programming, tax and interest expense, and franchise fees)

$$\text{Annual operating expense} = \underset{(1{,}526)^{a}}{10{,}390(\text{Miles})^{.71}} + \underset{(6.6)}{13(\text{Subscribers})}$$

$N = 17$ $R^2 = .94$ Standard error of the regression = 1,192.3

Equation 2: Estimated cable-system annual depreciation
(excluding equipment for local programming or access)

$$\text{Annual depreciation expense} = \underset{(295)^{a}}{536(\text{Miles})} + \underset{(190)}{975(\text{DMiles})} + \underset{(3.2)}{16(\text{Subscribers})}$$

$N = 17$ $R^2 = .99$ Standard error of the regression = 5,141.0

Note:
Subscribers = the number of basic-cable subscribers on the system.
Miles = the number of miles of operating plant in the system.
DMiles = the number of miles of operating plant: 0 if the system has a 12-channel capacity; 1 if the system has a 36-channel capacity

[a]standard error of the estimated coefficient.

12-channel system or $1,511 per mile for a 36-channel system with two-way capability. These figures represent book-value depreciation indexed to 1982 dollars with the producer price index for electrical equipment. The estimates are intended to reflect the cost of equipment purchased in 1982. Cable systems that have equipment purchased in earlier years, when prices were lower, will report lower annual depreciation expense, reflecting the diverse impact of inflation on the tax benefit of depreciation. Inflation indexing is, of course, an imperfect science, as revealed in the higher standard error of the coefficient estimating depreciation expense for older, 12-channel systems. The econometrically estimated depreciated expense is compatible with the estimates of cost as described in tables 3–1 and 3–2, which list major components of equipment.

Equations 1 and 2 in table 3–4 are designed to test the question proposed at the beginning of the chapter: From a cost perspective, is cable television a natural monopoly exhibiting declining average costs? These equations indicate that cable systems do exhibit declining average costs. An estimate of scale elasticities for cable systems appears in table 3–5.

Both depreciation and operating expense per channel seem to decline as the channel capacity of the system increases. If 36-channels are to be provided, it is cheaper for one firm to operate a 36-channel system than for three firms to operate 12-channel systems. Operating expense per mile of

Table 3-5
Scale Elasticities for Cable-Television Systems, from Equations 1 and 2 in Table 3-4

	Scale Elasticity for Market Penetration of:			
Miles of Cable	*20%*	*30%*	*40%*	*50%*
200	10.1	7.1	5.5	4.6
400	9.4	6.5	5.1	4.4

Notes:

Subscriber density: 100 households per mile of plant.

Channel capacity: 36.

Scale elasticity is defined as the percentage change in output (number of subscribers) divided by the accompanying percentage change in input (cost). A scale elasticity greater than one indicates the presence of economies of scale.

cable, illustrated in figure 3-4, declines as system length increases. If 900 miles of cable are to be installed in one geographic area, it would be cheaper to allow one firm to operate the 900 miles than to have three firms each operating 300 miles of cable.

Most dramatic is the declining cost per subscriber of providing service in an area given a cable system of fixed channel capacity and length. An industry rule of thumb has been that systems cannot be operated profitably when subscriber density falls below forty or fifty per mile. Figure 3-5, which approximates the patterns of costs expected for a public-utility or local telephone loop, is an illustration of the average total cost of providing service to a varying number of subscribers given a cable system 1,000 miles in length and having a 36-channel capacity. Average total costs decline as market penetration, or subscriber density, increases. As a result, it is cheaper for a single firm to provide service in a geographic area than to have two or more firms competing in the same area. Neglected in this conclusion are the direct and indirect costs of regulation that may be incurred if the cable system is to operate as a monopoly. Nevertheless, the notion of declining average costs provides the economic rational for the widespread practice of allowing cable-television systems to operate as franchised monopolies. From the perspective of cost, cable television is a natural geographic monopoly.

Franchise Fees

A cost specific to the cable industry is the franchise fee imposed on cable operators by franchising authorities. The fee, a percentage of the gross revenues legally limited to 3 percent, is intended to offset the cost of regula-

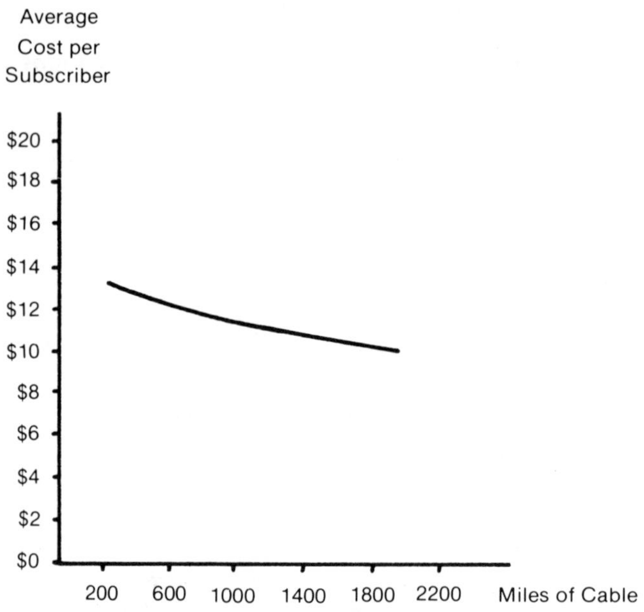

Figure 3-4. Average Total Costs for Cable Television Given Fixed Market Penetration

tion. Cable operators object to the fee, which they assert discriminates against their industry. In general, they have argued in favor of an amusement tax, which would fall equally on cable and all its potential competitors.

The Cost of Capital

Financing the growth of cable through the 1980s will require the investment of billions of dollars per year. In 1982 the industry retained earnings of over $1 billion for use in expansion. Outside sources of capital amounted to about $2 billion in debt and more than $500 million in equity, including limited partnerships and new stock issues. What is the cost of raising this capital? In economic terminology, what is a normal profit for a cable-television firm?

Of the two sources of capital, debt and equity, calculating the cost of dept inspires the least controversy. Interest rates for debt raised by the industry in 1982 were about 16 percent, up from about 10 percent in 1979. Since interest payments are tax deductible and the appropriate tax rate is

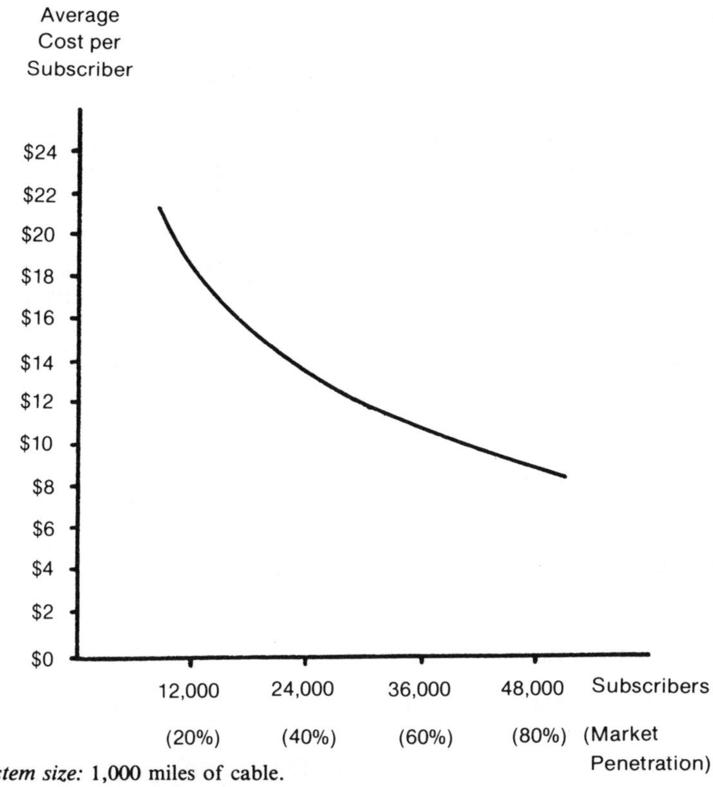

System size: 1,000 miles of cable.
Subscriber density: 100 households per mile of cable.

Figure 3–5. Average Total Cost for Cable Television Given Fixed Plant Size

about 48 percent, the after-tax cost of debt during this period ranged from about 4.8 to 7.7 percent. These figures represent nominal interest rates, unadjusted for inflation.

As an indication of the unusual conditions in the financial markets during recent years, consider values for the cost of debt in real terms. The real interest rate is approximately the nominal interest rate minus the inflation rate. In 1982, when cable companies were obtaining loans at 16 percent, the inflation rate was about 6 percent. The real rate of interest, then, was about 10 percent, something of a record in the market. During 1979, when cable operators obtained loans at 10 percent, the inflation rate, as measured by the GNP deflator, was about 9 percent. The implied real interest rate was about 1 percent. Part of the explanation for the lower interest rate during

1979 is that inflation was higher than anticipated. The reverse was true in 1982. Fluctuations in the real interest rate account for part of the business risk associated with using debt in the corporate financial structure.

Calculating the cost of equity is a problem that has generated thousands of pages of transcripts for public-utility rate hearings and professional-journal articles. Three popular methods of estimating the cost of equity are compared in this section.

A rule of thumb for the cost of equity is the ratio of the firm's after-tax earnings per share to its stock price, or *E/P*. For the major cable companies operating during the early 1980s, this measure of the cost of equity yields a value ranging from about 4 to 10 percent. The figure represents an after-tax rate. A formula thought to be more appropriate by financial analysts, especially for high growth stocks such as cable, is given by equation 3.1.

$$\textit{Cost of equity: } E(r) = \frac{D}{P} + G \qquad (3.1)$$

where $E(r)$ is the expected return on invested equity.

D is the firm's dividend.

P is the price per share of stock.

G is the expected growth rate.

Values of G have variously been estimated between 11 percent (Donaldson, Lufkin, and Jenrette 1979) and 20 percent (Merrill Lynch). Dividends for cable stocks are usually very low since most firms are using funds to finance expansion. The capital-asset pricing model (CAPM), which correlates the expected rate of return on an investment to a risk premium, is given by equation 3.2.

$$\textit{Capital-asset pricing model (CAPM): } E(r) = r_f + \beta(r_m - r_f) \qquad (3.2)$$

where $E(r)$ is the expected return on invested equity.

r_f is the risk-free rate of interest; for example, the interest rate on ninety-day treasury bills.

r_m is the return on the market portfolio, an investment containing the weighted average of all the investments available in the market.

β is the risk-adjustment factor for the form—the covariance of the return on the individual firm (r) and the market portfolio = $\text{Cov}(r_f, r_m)/\text{Var}(r_m)$.

The values for the parameters r_f and r_m, net of inflation, were estimated to be approximately 0 and 5 percent, respectively, during the last half of the 1970s (Merton 1980).

An approximation for the value of β (beta), a measure of the risk of the cable industry relative to the risk of the market in equation 3.2, is provided by the estimated β for Teleprompter, Inc.. A large corporation with investments almost exclusively in cable services, Teleprompter can be expected to reflect the risk of the industry. For the period 1974–1979, the β of Teleprompter, 1.65, suggests that cable television is a riskier investment than average. An average β is equal to one. The higher β for Teleprompter indicates that the expected return on equity for Teleprompter should be higher than average.

The expected after-tax cost of equity capital for cable, net of inflation, should be about 8.2 percent, according to the assumptions made in developing the CAPM equation. Allowing for an expected core inflation rate of about 8 percent, the nominal estimate becomes about 16 percent. This value falls into the middle of the 12- to 20-percent range suggested by the growth model (equation 3.1). Although these two models generate compatible results, the range of reasonable estimates is quite large.

During 1982 about one-third of the capital raised by the industry came from equity: joint ventures, new stock issues, and retained earnings. The overall industry average is somewhat lower, however. The debt-to-equity ratio of Teleprompter for the period reflecting the β of 1.65 was about one to one. Most of the major cable companies operating during the early 1980s enjoyed lower debt-to-equity ratios.

Informally combining these figures, which describe the cost of debt and equity, yields an average cost of capital in the range of about 4 to 8 percent, net of inflation. The wide range of this estimate represents the uncertainty associated with quantifying the economic concept of normal profits.

Summary

Cable systems appear to exhibit declining average total costs in terms of the number of channels provided, the size of the geographic area covered, and the number of subscribers served. From the perspective of costs, cable television systems are a natural geographic monopoly.

Construction costs for cable systems vary depending on channel capacity, system length, subscriber density, production facility, interactive capability, and the nature of the local terrain (which may require underground or aerial distribution plant). Among the technological uncertainties in the industry's future is the length of time that will be required to make fiber optics cost-effective for use in cable systems.

Appendix 3A Estimation Procedure for Statistical Cost Functions

The cost equations in table 3–4 were estimated simultaneously using the computer program Time Series Processor (TSP). TSP is designed to handle small cross-sectional problems and relies on the Gauss nonlinear least-squares estimator for simultaneous, nonlinear-equation estimation. The equations, estimated simultaneously to eliminate bias resulting from the interdependence of disturbance terms, are in the form of 3A.1 and 3A.2.

$$\frac{\text{Op-Expense}}{\text{Subs}^{.5}} = \frac{a_0}{\text{Subs}^{.5}} + \frac{a_1(\text{Miles})^{a_2}}{\text{Subs}^{.5}} + \frac{a_3(\text{Subs})}{\text{Subs}^{.5}} \tag{3A.1}$$

$$\frac{\text{Dep-Expense}}{\text{Miles}^{.5}} = \frac{b_0}{\text{Miles}^{.5}} + \frac{b_1(\text{Miles})}{\text{Miles}^{.5}} + \frac{b_2(\text{DMiles})}{\text{Miles}^{.5}} + \frac{b_3(\text{Subs})}{\text{Miles}^{.5}} \tag{3A.2}$$

Constant terms (a_0 and b_0) in both equations were constrained to be nonnegative. Based on the test for heteroskedasticity developed by Glejser (1969), corrections for heteroskedastic disturbances were made to the equations. The absolute values of the disturbance terms were regressed against each independent variable. In the equation for operating expense, disturbance terms were significantly correlated with the number of subscribers served. In the equation for depreciation expense, the disturbance terms were significantly correlated with the length of the system, the miles variable. In both cases, the disturbance terms were approximately proportional to the square root of the relevant variable. A search of alternative values for the exponent at 0.1 increments, over the range 0 to 2 confirmed that heteroskedasticity is minimized when the number of subscribers (equation 3A.1) and number of miles (equation 3A.2) are raised to the 0.5 power in the denominator of the equations. Both equations passed the other standard tests, described in appendix 5B, used to verify that disturbance terms are well behaved.

4 The Demand for Cable Services

What is the nature of the market for cable services? Who is likely to subscribe and at what price? How do related services such as broadcast television, motion-picture theaters, or subscription television (STV) affect the demand for cable?

Corporations and municipalities must answer these questions when deciding what services to offer and what prices to charge. The question of related services also raises a more general issue about the desirability of price regulation. Unless related services pose significant competition, the market cannot be expected to ensure reasonable cable-television prices when the cable system is operating as a franchised monopoly.

This chapter explores the nature of the cable-television market, using an econometric model. The model is based on data reported in 1980 by ninety-eight urban cable systems operating throughout the United States and on a large market survey conducted in Philadelphia during the period 1980–1982.

The several equations that make up the econometric model are summarized in appendix 4A. Also summarized is a discussion of the statistical problems involved in estimating the equations. About 70 percent of the variation in the demand for basic cable and about 30 percent of the variation in the demand for pay cable can be explained by the model. By comparison, the equations included in the previous chapter accounted for more than 90 percent of the variation in costs among the cable systems in the sample. As is often true, the model of demand contains more uncertainty than the model of cost. Part of the usefulness of the econometric model is that it provides an estimate of this uncertainty. Additionally, several of the most important determinants of demand, especially price, seem to be well explained by the model.

The measures of demand most frequently used in the industry are the market penetration (MP) of basic cable and the system penetration (SP) of pay cable. Defined in equations 4.1 and 4.2, MP and SP measure the number of subscribers as a percentage of households in the market.

$$\text{MP} = \frac{\text{Number of households subscribing to the basic-cable system}}{\text{Number of households passed by cable}} \qquad (4.1)$$

$$\text{SP} = \frac{\text{Number of households subscribing to a pay-cable service}}{\text{Number of households subscribing to the basic-cable system}} \tag{4.2}$$

Each cable system serves a well-defined market area; but, because no two cable systems serve exactly the same-sized market, comparisons based solely on the number of subscribers can be misleading. A small system may be doing very well to attract 1,000 subscribers, whereas for a large system the number would represent a dismal failure. Penetration, based on a percentage, is a meaningful measure of demand for a system of any size; demand is described in terms of market potential. The demand for pay cable is defined as system penetration because at present, for most systems, households wanting to subscribe to pay-cable services must also subscribe to basic cable. Pay cable is available only to basic cable subscribers. In the future this requirement may be relaxed.

In general, the determinants of demand for cable services include the socioeconomic characteristics of the consumers in the market, the geographic character of the city served by the system, the availability of broadcast television, the numbers and types of channels offered by cable, and the price of cable services. A measure that describes the effect of these determinants is the elasticity of demand.

The Elasticity of Demand

Equation 4.3 defines a general form for the elasticity of demand.

$$\eta = \frac{\%\Delta P}{\%\Delta D} \quad \text{or, more formally} \quad \frac{\partial P}{\partial D} \cdot \frac{D}{P} \tag{4.3}$$

where η is the elasticity of demand.

Δ is change.

P is *penetration,* either the market penetration of basic cable (MP) or the system penetration of pay cable (SP).

D is one of the *determinants* of demand—for example, the price level.

A convenient form is the arc elasticity of demand, which avoids any confusion about what values should be used in calculating percentage change.

$$\text{Arc elasticity} = \left[\frac{(P_1 - P_2)}{(P_1 + P_2)}\right] \left[\frac{(D_1 + D_2)}{(D_1 - D_2)}\right]$$

where $(P_1 - P_2)$ represents the change in market penetration and $(D_1 - D_2)$ represents the change in the level of the determinant.

The elasticity of demand measures the sensitivity of demand to changes in the level of demand determinants. An elasticity of two, for example, indicates that a 1-percent increase in the level of a determinant results in a 2-percent increase in demand. An elasticity of zero indicates that no change in demand would result from a change in the level of the determinant.

The elasticity of demand is probably not the same at all levels of market penetration. For example, a 10-percent change in the subscription price for a cable service may be expected to have a larger percentage impact on demand for a system at a low penetration level than for one at a high penetration level. Statistical tests, described in appendix 4A, confirmed this assumption. To capture the effect, several of the elasticities were estimated and are reported in the form of equation 4.4 for basic cable and equation 4.5 for pay cable.

$$\eta_b = [1 - \text{MP}]C \tag{4.4}$$

$$\eta_p = [1 - \text{SP}]C \tag{4.5}$$

where η_b is the elasticity of demand for basic cable.

η_p is the elasticity of demand for pay cable.

MP is the market penetration of basic cable, taking values from 0 to 1.0.

SP is the system penetration of pay cable, taking values from 0 to 1.0.

C is an estimated constant term.

This form of the demand elasticity results from the use of the popular logit model, discussed in appendix 4A, which provides a somewhat better statistical fit of the data than do simpler forms such as the linear or log-linear models.

Characteristics of the Consumer: Socioeconomic Variables

Among the most commonly reported demand elasticities, the income elasticity describes how consumers adjust their expenditures to different levels of income. It also provides the economic distinction between an essential and a nonessential good. Recall that one of the arguments made in favor of allowing the deregulation of prices was that cable television represents a luxury, or a nonessential good. Although this notion has little economic theory to support it, intuition may suggest that the prices of essential goods should be scrutinized more closely than those of nonessential goods.

A nonessential good is said to have an income elasticity greater than one, indicating that consumers' expenditures on the good increase more than proportionately with income. Members of high-income groups will spend a larger share of their income on a nonessential good than will members of low-income groups. Historically, cable operators have often preferred to construct systems in high-income neighborhoods, assuming that demand would be strongest there.

The results of the econometric model, reported in table 4–1, indicate that pay cable may be a nonessential good by the economic definition. For system-penetration rates below 60 percent, the estimated income elasticity is greater than one. High-income neighborhoods appear to be very good target markets for pay cable. The reported income elasticity for basic cable, however, is negative. In the cities that made up the national data, low-income families spent more dollars on basic cable than did middle-income families. On the basis of this evidence, it seems that cable operators may have undervalued a potentially important part of the market. Low-income neighborhoods appear to be good target markets for basic cable. It also means that special taxes on cable, such as municipal franchise fees, fall heavily on low-income groups.

Should the negative income elasticity be expected? Especially in urban areas, which the data represent, low-income families may find that basic cable provides an inexpensive substitute for the other kinds of entertainment that the city offers.

Evidence that appears to contradict the measured negative income elasticity can be found in several studies conducted during the 1970s for the FCC (Charles River Associates 1978a, b; Park 1971; Schink and Thanawala 1978) and the results of the market survey for Philadelphia described in appendix 4A. The FCC-sponsored studies, however, include data on many small-market cable systems and thus would not have measured the negative income elasticity if it is an urban phenomenon. The market survey of Philadelphia, gathered by students from a private university who sometimes surveyed family and neighbors, represents a higher-than-average income group. The survey results suggest that at an income level well above the national average, the income elasticity becomes positive. Weak statistical evidence for this relationship was found in the national data.

Often the demand elasticity of income has the same sign, positive or negative, as the effect of the level of education on demand. Higher incomes

Table 4–1
Demand Elasticities for Socioeconomic Variables

Variable	*Demand Elasticity*[a] *for Basic Cable*	*Demand Elasticity*[a] *for Pay Cable*
Income, per capita[b]	−1.2[1 − MP] (0.4)[c]	2.5[1 − SP] (1.0)[c]
Index of Home Equipment, the combined percentage of occupied housing units with air conditioning, home food freezer, telephone, and one or more automobiles (average 218 percent; highest possible 400)	0.2[1 − MP] (0.09)	ns
Education, percentage of population having completed a high-school education (average 56.9 percent)	0.5[1 − MP] (0.03)	ns
Variables tested, but not found to be statistically significant		
Income distribution		
Population, percentage aged 65 or older		
Population, rate of change		

Note: ns = not significant.

[a]Values for MP, basic-cable market penetration, and SP, pay-cable system penetration, are expressed in decimal form. For example, an MP of 0.3 represents a 30-percent market penetration.

[b]For the income of $5,291–$14,377 per capita in 1982 dollars.

[c]Values in parentheses are the standard error of the estimate.

are usually associated with higher levels of education. As reported in table 4–1, however, populations having a higher percentage of high-school graduates displayed a stronger demand for basic-cable service. The effect of education on the demand for pay cable was not statistically measureable.

Perhaps the most unfamiliar determinant of demand included in the table is the Index of Home Equipment. Compiled by the Census Bureau, the index is the sum of the percentages of households that have air conditioning, a home food freezer, a telephone, and one or more automobiles. The maximum value of the index is 400. Of course, automobile market penetration, included in the index, seems an unlikely measure of potential cable-television demand. For that matter, what do air conditioners or home food freezers have to do with cable television?

The Index of Home Equipment does provide some useful general information about the pattern of consumer purchasing decisions. As reported in table 4–1, households with more home equipment are more likely to purchase cable-television services. The common denominator for most of these purchase decisions is that they improve the quality of the household environment. Many studies have revealed a strong neighborhood effect for this kind of purchase decision.

Another advantage of the index is its availability. The Census Bureau has gone to some expense to make this information accessible in the *City and County Data Book.* Reliable purchase information for other indicators of potential cable-television demand, such as expenditures on videotape equipment, is not readily available and can be expensive to acquire. Weighing the cost against the likely value of additional information constitutes a fundamental problem encountered in market research.

Table 4–1 lists four variables that were tested but found to contribute no statistically measurable information to the econometric model. These variables were considered likely determinants of cable demand. It has been suggested, for example, that people aged 65 and older are a good target market for cable service since they may have more leisure time available. Also, differences in buying behavior are often evidenced among ethnic groups. Population rate of change was found to have some impact on cable demand in an earlier study (Charles River Associates 1978a); transient households seemed less likely to purchase cable.

The Influence of Geography

The demand for cable television is, on average, about 20 percent lower in Kansas than in California or Pennsylvania. Cable systems in the West North Central states, such as Kansas, tend to attract fewer viewers than systems in other parts of the country. In the New England and South Atlantic states, demand tends to be higher than average. Estimates for the variation in penetration rates among geographical regions of the country appear in table 4–2. To some extent these variations reflect cultural differences. Television may be less a part of the Kansas life-style than of the California life-style. The differences might also be attributable to the availability of related services. The higher-than-average demand for pay cable in the Mountain states may reflect a lack of alternative forms of entertainment.

A more direct effect of geography is the extent to which local terrain interferes with broadcast-signal reception. Demand for cable is higher where reception is poor. In fact, the single most important determinant of cable demand is the availability of broadcast signals. Most of the cable systems in operation serve as a community antenna used to improve broadcast-signal reception. It was, after all, the effect of mountainous terrain that inspired the first cable systems. The measured impact of terrain, described in table 4–2, indicates that a system that would achieve a 20- or 30-percent penetration rate where the terrain was perfectly flat would achieve a penetration rate of over 90 percent in the most mountainous terrain.

Table 4–2
The Impact of Geography on the Demand for Cable

Location of the Cable System	*Percentage Difference from Average Basic-Cable Market Penetration (MP)*[a]	*Percentage Difference from Average Pay-Cable System Penetration (SP)*[a]
Mountain states, including: Arizona, Colorado, Idaho, Montana, New Mexico, Nevada, Utah	ns	0.2[1 − SP] (0.6)[b]
New England states, including: Connecticut, Maine, Massachusetts, New Hampshire, Rhode Island, Vermont	0.3[1 − MP] (0.3)	ns
South Atlantic states, including: Delaware, Florida, Georgia, Maryland, North Carolina, South Carolina, West Virginia	0.4[1 − MP] (.14)	ns
West North Central states, including: Iowa, Kansas, Minnesota, Nebraska, North Dakota, South Dakota	−0.45[1 − MP] (.19)	ns
Terrain		
The value squared of a Geologic Survey Scale describing the local terrain; ranging from 1 for flat terrain to 20 for highly mountanous	0.015[1 − MP] (.009)	ns
Variables tested but not found to be statistically significant		
Location in the East North Central, East South Central, Middle Atlantic, or Pacific states		
Percentage daily hours of sunshine		
Population density		
Television market rank		

Note: ns = not significant.

[a]Percentages are expressed in decimal form; that is, 0.3 represents 30 percent. MP and SP are also expressed as decimals. According to the estimates, a system operating with an MP of 0.55 in California would achieve, on average, an MP of about 0.45 in Kansas.

[b]Values in parenthesis are the standard error of the estimate.

Another geographic characteristic of crucial importance to the development of the communications industry is population density. For example, a rule of thumb in the cable industry has been that population density must be high enough to provide about forty subscribers per mile, or systems will not be profitable. As indicated in table 4–2, population density does not contribute to the demand model that forecasts market penetration; for a given level of market penetration, however, the number of subscribers per mile is greater where population density is higher. This relationship has encouraged the widespread interest among private companies in urban cable systems, which, serving the market areas of highest population density, are potentially the most lucrative.

Table 4–3
The Change in the Demand for Basic Cable over Time

Age of the Cable System	*Demand Elasticity*[a] *for Basic Cable*
Number of months the system has been in operation	0.24[1 − MP] (.06)[b]

Example[c]

Year	*Basic-Cable Market Penetration*	*Year*	*Basic-Cable Market Penetration*
1	.354	6	.457
2	.393	7	.466
3	.416	8	.474
4	.433	9	.481
5	.446	10	.487

Note: Most of the system penetration for each pay-cable service is achieved within the first year of availability. The total demand for pay cable, however, increases significantly as the number of services increases.

[a]The elasticity of demand with respect to time is the percentage change in basic-cable market penetration divided by the yearly percentage change in the number of months the system has been in operation.

[b]Standard error of the estimate.

[c]The example is based on a 40-channel system operating in the Northeast franchise of Philadelphia, as described in appendix 5B.

The Change in Demand over Time

Table 4–3 presents an example of the growth in demand for cable that represents the historical pattern for individual systems. It shows that basic cable achieves most of its market penetration in the first few years of operation; the market matures rapidly. This growth rate is abstracted from any change in the nature of the industry, such as service innovation or competition. Although the timing may be different, many products undergo a similar diffusion process when introduced into a market. Unsure about the potential benefits, some households will decide to become subscribers on the basis of their neighbors' experience.

A pay-cable channel achieves most of its system penetration in the first year of availability. This result suggests that households make the decision to subscribe to basic and pay cable at about the same time. For pay-cable services added sometime after the system becomes operational, the decision to subscribe is most often made when services become available. Of course, the most important force driving cable growth has been the expansion of cable services.

Cable Services

By far the most important determinant of demand for basic cable is the programming service it provides compared with the local broadcast market. The equation in footnote a to table 4–4, which summarizes the comparative benefits of cable service, is based on the ratio of the number of signals carried by cable to the number of stations available over the air. The cable system is at a comparative advantage when it offers more signals than are available over the air, when the ratio is greater than one. Of course, a signal is only as important as the programming it carries. Also relevant is the technical quality of the signal.

Table 4–4
The Demand Impact of Cable-System Services and Characteristics

Broadcast and Imported Signals Carried on Basic-Cable	*Demand Elasticity of Comparative Service*[a] *for Basic-Cable*	*Demand Elasticity of Comparative Service*[a] *for a Pay-Cable Channel*
Network: A signal broadcast from the nearest local affiliate of ABC, CBS, or NBC.	1.4[1 – MP] (0.29)[b]	–0.9[1 – SP] (0.38)
Duplicate network: In local markets served by more than one broadcast station affiliated with the same network, the more distant network affiliated signal or signals.	0.2[1 – MP] (0.09)	ns
Independent: A signal from a satellite-distributed network or from a broadcast station not affiliated with ABC, CBS, NBC, or PBS.	0.4[1 – MP] (0.1)	ns
Locally Originated Programming		*Percentage Increase in the Demand for Basic-Cable for Each Additional Channel of Local Origination*
Number of channels dedicated to public access		0.06[1 – MP] (0.03)
Number of channels dedicated to locally produced programming		0.066[1 – MP] (0.08)

Table 4–4 continued

Variables Tested, but Not Found To Be Statistically Significant
Installation fee (most installation fees are quite low, about $20, and vary only a little among systems)
Length of the system
Number of motion picture theaters near the cable franchise area
Number of pay-television channels carried by the system
Number of radio stations carried by the system
Two-way capability

[a]Comparative service, X_i, is:

$$X_i = \text{Log}\left[\frac{1 + S_i}{1 + \Sigma_{j=1}^{n} P_{ij}(R)}\right]$$

where i = 1, 2, 3 types of programming; network, duplicate network, independent.

X_i = the ratio of channels of type i programming offered by the cable system as compared to the number available over the air, adjusted by the probability of good reception.

S_i = the number of channels of type i carried on the cable system.

n = the number of channels of type i available over the air as broadcast stations.

$P_{ij}(R)$ = the percentage of viewers who receive satisfactory reception of the over-the-air signal for station j or type i.

[b]Values in parentheses are the standard error of the estimates.

[c]For example, a system operating at 25 percent market penetration may expect about a 4.5-percent increase in the number of basic-cable subscribers by adding a public-access channel. Note that measured impact of locally produced programming is not significant at the 95-percent confidence interval.

In the equation, four broad categories of programming are identified on the basis of station affiliation or ownership: network, duplicate network, independent, and public. *Network* stations are those owned by or affiliated with ABC, CBS, or NBC. In markets where a network is represented by more than one station, the more distant network stations are designated *duplicate network* stations. Sources of *independent* programming include unaffiliated broadcast stations and the satellite networks such as Nickelodeon and Cable News Network. Educational stations, or those affiliated with PBS, make up the public category; but the demand impact of educational stations, which historically have attracted small audiences, was too small to be measured statistically.

The other aspect of comparative service, signal quality, was captured by estimating the percentage of viewers in the local market who receive an adequate over-the-air broadcast signal. Measurements of broadcast-signal strength conducted by the FCC provide two benchmarks ideally suited to this definition of signal quality: the A and B contours.

The A contour describes an approximately circular boundary around the broadcast station within which the median observer receives a satisfactory signal at least 90 percent of the time for 90 percent of the locations. At the B contour the median observer receives a satisfactory signal at least 90 percent of the time for 50 percent of the locations. For a city located on the B contour of a local broadcast station, half the households receive an adequate signal and half do not.

The number of miles from a station's broadcast tower to the A or B contour depends on the power of the station's transmitter and the height of the broadcast tower. For most stations the B contour is forty to seventy miles from the broadcast tower. Maps of the A and B contours for all operating broadcast stations are printed in the *Television Factbook* (1980). A method of approximating the percentage of viewers who receive adequate signal reception for locations between the A and B contours is discussed in appendix 4A.

The results of the econometric model, summarized in table 4–4, indicate that basic-cable market penetration is increased significantly by improving the signal quality of a network station or by importing a network signal into a market not served by a network station. The system penetration of pay cable, however, declines by almost two-thirds as much as the basic-cable penetration increases.

For example, imagine that a cable system has been constructed in a city with two nearby network stations that are adequately received by everyone, and one distant network station that is adequately received by only 50 percent of the viewers. The cable system provides, among other services, good reception of the distant network station. By how much will improved reception of the network station increase basic-cable market penetration? If the cable system offered no network station comparative service advantage, the value of the equation in table 4–4 would be one. Counting the improved reception of the distant network station, the cable system offers 3 network stations and the local market offers only 2.5 (half the viewers do not receive a good off-the-air signal for the distant network station); the value of the equation is 1.2 (3 divided by 2.5). The measured comparative service has increased from 1 to 1.2, about 20 percent.

To translate this figure into a change in basic-cable market penetration, it is only necessary to multiply by the value of the demand elasticity. For a system with a market penetration of 0.3, or 30 percent, the network demand elasticity is about one (from table 4–4: 1.4[1 − .03] = 0.98). An elasticity of one means that a 20-percent increase in comparative service should give rise to about a 20-percent increase in the number of basic-cable subscribers. Basic-cable market penetration would therefore increase from 30 to 36 percent. The system penetration of pay cable, however, will fall. If pay-cable system penetration is also about 0.3, or 30 percent, it will fall by about 12.5 percent as a result of the improved network-station reception (from table 4–4: $0.2[-0.9[1 - 0.3]] = -0.1265$). The new level of system penetration would be about 0.26, or 26 percent. This does not mean that the number of pay-cable subscribers has declined. Pay-cable system penetration—the percentage of basic-cable subscribers who also subscribe to pay-cable—declines, but the number of pay-cable subscribers actually increases.

To understand this relationship, imagine that there are 1,000 households in the market area. At first, the basic-cable market penetration was 30 percent, or 300 subscribers. The pay-cable system penetration was also 30 percent, or 90 subscribers (30 percent of the basic-cable subscribers). Basic-cable market penetration then increased to 36 percent for 360 subscribers, 60 more than before. Pay-cable system penetration decreased to 26 percent for about 94 subscribers, but this represents an increase of 4 subscribers. Of the 60 new basic-cable subscribers in our example, 4 decided to subscribe to pay cable also. Increasing the market penetration of basic cable also increased, though by a smaller amount, the number of pay-cable subscribers.

As the example illustrates, cable operators may significantly increase basic-cable market penetration by providing improved reception of network-affiliated stations. Through the early 1980s, improved reception of broadcast signals still represented the most important aspect of demand for cable. Although the industry is developing as a purveyor of new information services, it still operates primarily as a community antenna.

Much of cable's future rides on the demand for additional, independently produced signals. The historical impact of providing signals from broadcast independents and satellite networks is measured in table 4–4 by the demand elasticity for independent signals. Since this measurement is based on historical information for systems offering at most ten or fifteen independent signals, some caution must be exercised in extrapolating too far into the future for large cable systems offering numerous channels.

As an example of the demand elasticity for independent signals, consider a cable system at 50-percent basic-cable market penetration that distributes three local independent stations. Market penetration would increase to about 54 percent if three additional independent signals were distributed. The difference amounts to about an 8-percent increase in the number of subscribers.

Of course, a change in service level may also work against the cable operator. For example, if three low-power television stations began broadcasting conventional programming in the area and were not distributed by the cable system, demand for basic-cable would decrease to a level of about 46-percent market penetration.

Some of the local programming that planners anticipate will be distributed by low-power television stations is already available on many cable systems. The prospect of locally originated programming has inspired optimistic visions of neighborhood studios busily producing local theater, politics, and sports. Most systems accommodate local origination; but, as reported in table 4–4, demand is limited. For a basic-cable system at 50-percent market penetration, an additional channel of local origination can be expected to increase market penetration to only about 51.5 percent, a 3-percent increase in the number of subscribers. In fact, the difference is not statistically significant at the 95-percent confidence level for locally produced programming. For public-access programming, however, the difference is significant.

For many cable systems, the increased demand expected from local origination will not generate enough revenue to offset the production costs. Left to themselves, profit-maximizing cable operators can be expected to fill channels with signals from broadcast stations or satellite networks, or with pay-television programming. Unless advertising sales can be used to augment revenue, local origination is not profitable. That many systems support local origination must be attributed partly to regulatory intervention.

Regulatory intervention also seems to be responsible for the presence of two-way capability on many systems constructed during the 1970s. In the early part of the decade, experiments with two-way entertainment services failed to demonstrate that any significant demand existed. With long-term goals in mind, the FCC required that systems constructed during the mid-1970s include components allowing for two-way capability. According to the historical data summarized in table 4–4, no significant demand for the available two-way entertainment services existed prior to the early 1980s. Of course, historical data may underestimate future demand for innovative services. A survey revealed a potential market for interactive computer games (see table 4–5). The potential demand is about twice as great as the demand for public-access programming.

The cable industry has recently focused attention on the development of two-way nonentertainment services: home-security systems, interactive shopping, and data transmission. As indicated by this attention, two-way nonentertainment services represent a large potential market.

Historically, the communications link required for most two-way nonentertainment services has been provided by telephone, which is still locally rate-regulated. When competing against a regulated firm, an unregulated firm will tend to serve only the most lucrative part of the market, leaving the regulated firm with only the less-profitable or unprofitable part of the market. This pattern of behavior, commonly characterized as cream-skimming, would provide a large revenue source to the cable industry at the expense of the telephone industry. The prospect has prompted legislative proposals that would restrict the cable industry from entering nonentertainment markets. Until Congress decides on the extent to which cable will be allowed to compete, any estimate of potential revenue for cable systems from nonentertainment must contain a great deal of uncertainty.

Table 4-5
Programming Preferences

Programming	*Among Most Preferred for Percentage of Households*[a]
Current movies	70.1
Sports	41.1
News	37.6
Old movies	30.9
Adult movies	30.6
Musical	28.3
Computer games	23.6
Local government	11.8
Local entertainment	10.5
Public access	8.6
Religious	6.7
Ethnic	6.4
Children's	0.3

[a]Percentage of adult respondents interviewed who indicated that the programming was among their most preferred. The survey was conducted in 1982 for 330 households in Philadelphia.

Among the recent changes in the delivery of entertainment services associated with cable has been the emergence of satellite-distributed networks offering specialized programming such as news, sports, or religion. Broadcast stations have historically offered a mix of these programs. The distinction is not fully measured in table 4–4, which summarizes the demand elasticity for additional channels. Not enough data were available to separate demand completely by program content. Results of the market survey, reported in table 4–5, offers some additional information.

The strong demand measured for current movies and sports is reflected in the market by the number of pay-television networks offering this kind of programming. According to the survey, the demand for adult movies, the so-called blue channels, appears to be about three times as strong as for local programming.

The relatively weak demand measured with the market survey for local programming conforms to the historical evidence. A strong demand for educational programming measured by the market survey appears to contradict the historical evidence. Among students at Drexel University who conducted interviews for the survey, it was theorized that respondents overstated their demand for educational programming in order to impress the interviewer.

Throughout eight successive surveys, each with a sample in excess of 300, respondents consistently overestimated by about twofold the amount of money that historical evidence in the market demonstrated they were willing to spend on cable services. As market research often reveals—but as researchers sometimes fail to take into account—expressed desires often exceed the reality of the pocketbook.

The market survey was also designed to address the competitive potential of noncable pay-television distribution such as subscription television (STV). The results indicate that about 10 percent of the households would be expected to subscribe to STV where cable was not available and less than 5 percent where cable was available. According to a market survey conducted for the National Cable Television Association, STV can be expected to draw about 3 percent of the pay-television market from pay cable where a cable system preceded the STV station. Where STV preceded cable, as much as 14 percent of the pay-television market may be lost to STV.

Although much has been written about the long-run competition between cable and motion-picture theaters, the availability of motion-picture theaters did not measurably influence demand for cable. Most of the variability in the demand for cable results from the level of service offered by the cable system itself. These results support the wisdom of many financial analysts who expect little close competition from new noncable entertainment services, except where cable operates at a significant price disadvantage.

Cable-System Prices

The lack of apparent close substitutes for basic cable suggests that demand should be relatively insensitive to changes in price. Households are inclined to endure a price increase when there are few substitutes available in the form of alternative services. As reported in table 4–6, the price elasticity of demand for basic cable was measured to be between -1 and 0 over the normal range of prices. The negative value confirms, as expected, that market penetration and price are inversely related. Demand decreases (increases) as price increases (decreases). Since the negative sign can almost always be expected, the price elasticity of demand is typically discussed in terms of absolute value. An absolute value of less than 1 indicates, also as expected, that the demand for basic cable is relatively insensitive to changes in price; in economic terms, demand is inelastic. Goods such as food, clothing, and transportation also generally exhibit an inelastic demand.

Demand for basic cable becomes more elastic—more price sensitive—as prices rise and market penetration declines. Maximum total basic-cable revenue will be reached when the price elasticity of demand equals 1 in

Table 4–6
Price Elasticities for Cable Services

Type of Elasticity	*Demand Elasticity for Basic-Cable Market Penetration (MP)*	*Demand Elasticity for Pay-Cable System Penetration (SP)*
Own-price[a]	−1.0(1 − MP) (.37)[d]	−1.7(1 − SP) (.93)
Cross-price[b]	ns	ns
Super cross-price[c]	ns	−1.0(1 − MP)SP

Notes: ns = not significant.
Δ = change.
P_b = the price of basic cable.
P_p = the price of pay cable.

[a]

	(for basic cable)	(for pay cable)
Own-price elasticity:	$\frac{\%\Delta MP}{\%\Delta P_b}$	$\frac{\%\Delta SP}{\%\Delta P_p}$

[b]

	(for basic cable with respect to the price of pay cable)	(for pay cable with respect to the price of basic cable)
Cross-price elasticity	$\frac{\%\Delta MP}{\%\Delta P_p}$	$\frac{\%\Delta SP}{\%\Delta P_b}$

[c]Super cross-price elasticity (for pay cable with respect to the price of basic cable) $\left[\frac{\%\Delta SP}{\%\Delta P_b} + \frac{\%\Delta MP}{\%\Delta P_b}\right] SP$

[d]The value in parenthesis is the standard error of the estimate.

Example

Basic-Cable Price	*Basic-Cable Market Penetration*
$14	23%
12	26
10	30
8	35
6	41
4	50

absolute value. Therefore, cable operators could increase revenues by increasing basic-cable prices. Presumably only regulation, the fear of regulation, or the fear of new competition has kept prices as low as they currently are for systems that offer only basic-cable service.

How high would prices rise if these controls were eliminated? Historical data cannot answer this question with precision. Over the range of market penetrations included in the sample, about 25–80 percent, demand is inelastic as measured by the econometric model. According to the model, demand becomes elastic at market-penetration levels of nearly zero. Common sense, however, dictates that prices high enough to exclude all but 1 or 2 percent of the market probably do not represent the best strategy. The contradiction between the model and common sense for these extreme values serves to illustrate the danger of extrapolating far beyond the sample on which assumptions have been constructed. Over the relevant range, however, the model yields reliable evidence to suggest that cable operators offering only basic cable could increase short-run profits by increasing prices.

Besides the monthly subscription price, cable subscribers usually also pay an installation fee, averaging about $20. That price level does not reflect the full cost of installation. Cable operators seem to be choosing a price high enough to discourage subscribers from connecting and disconnecting service at whimsically frequent intervals, but low enough so that it has no significant negative influence on demand. Since consumers are usually more sensitive to lump-sum prices, cable operators have wisely prorated part of the actual cost of installation into the basic-cable subscriber fee. Of course, the cost associated with disconnection must also be prorated into the monthly subscriber fee. Although it may be economically justifiable, few firms attempt to bill subscribers for the cost of disconnecting service.

An aspect of basic-cable demand not included in the notion of market penetration is the extent to which subscribers desire multiple connections to the system. Households having more than one television set or commercial establishments such as motels are likely candidates for multiple connections. About 22 percent of the households paid an average of $2.40 per month for at least one additional connection.

A service included as part of basic cable on some systems, but offered on others at a monthly fee of $1 or $2, is cable radio. The determinants of demand for cable radio are similar to those for basic cable. Demand is higher for cable radio where there are few radio stations in the local market and where the cable system offers signals imported from outside the market. At present, demand for the service is so limited that price elasticities are not measurable.

The market provides better information about the demand for pay-cable services such as HBO. As might be expected, the demand for pay cable is more elastic than the demand for basic cable. One reason is that pay-cable rates are not regulated. Firms would not voluntarily choose to set prices in the inelastic range where price increases lead to higher profits. Also, there are close substitutes for pay cable. In a few markets, direct competition exists among pay-television services. Demand becomes more elastic as competition intensifies. For pay-cable system-penetration levels of less than about 40 percent, the price elasticity of demand is greater than 1 in absolute value—demand is elastic. When demand is elastic, total revenue can be increased by lowering prices.

An example of the effect of price on the demand for cable service appears in figure 4–1. Based on market conditions in Philadelphia, as described in appendix 5B, the demand line schedules the expected cable penetration at each price level. As the price of service approaches 0, penetration approaches 100 percent. At the other end of the price range, a small percentage of the market is willing to pay $20 per month or more for cable service.

In markets where there are fewer broadcast signals or where reception is poor, the demand line will be shifted to the right. Increasing the number of services offered by basic cable will also shift demand to the right. Another important aspect of the market is the demand interdependence of cable services. Changing the price of basic cable affects the demand for pay cable.

Cross-Price Effects

Over the normal range of prices, basic-cable system penetration seems to be insensitive to pay-cable prices. One explanation for the lack of measurable direct cross-price effects is the curious nature of the relationship between basic and pay cable. These are potentially substitute goods. If the consumer could choose between the two, high pay-cable prices would make basic cable appear to be a better buy. This relationship would be one of substitutability; the cross-price elasticity would be positive. Raising pay-cable prices would increase the level of basic-cable demand, or vice versa. The market survey revealed that as many as 15 percent of the pay-cable subscribers would not subscribe to basic cable if that choice were available. On almost all cable systems, however, households must subscribe to basic cable in order to subscribe to pay cable. Enforcement of this technological requirement changes the nature of the relationship from one of substitutability to one of complementarity. Lowering pay-cable prices should encourage individuals to subscribe to basic cable for the privilege of having access to inexpensive pay cable. Of course, the price change must be large enough so that individuals perceive a benefit. Historically there has been little cross-sectional variation in pay-cable prices. As a result, the market does not provide much information about the nature of this relationship.

Better information was available for the complementary effect of basic-cable prices on pay-cable demand, measured by the super cross-price elasticity in table 4–6. The super cross-price elasticity indicates that for a cable

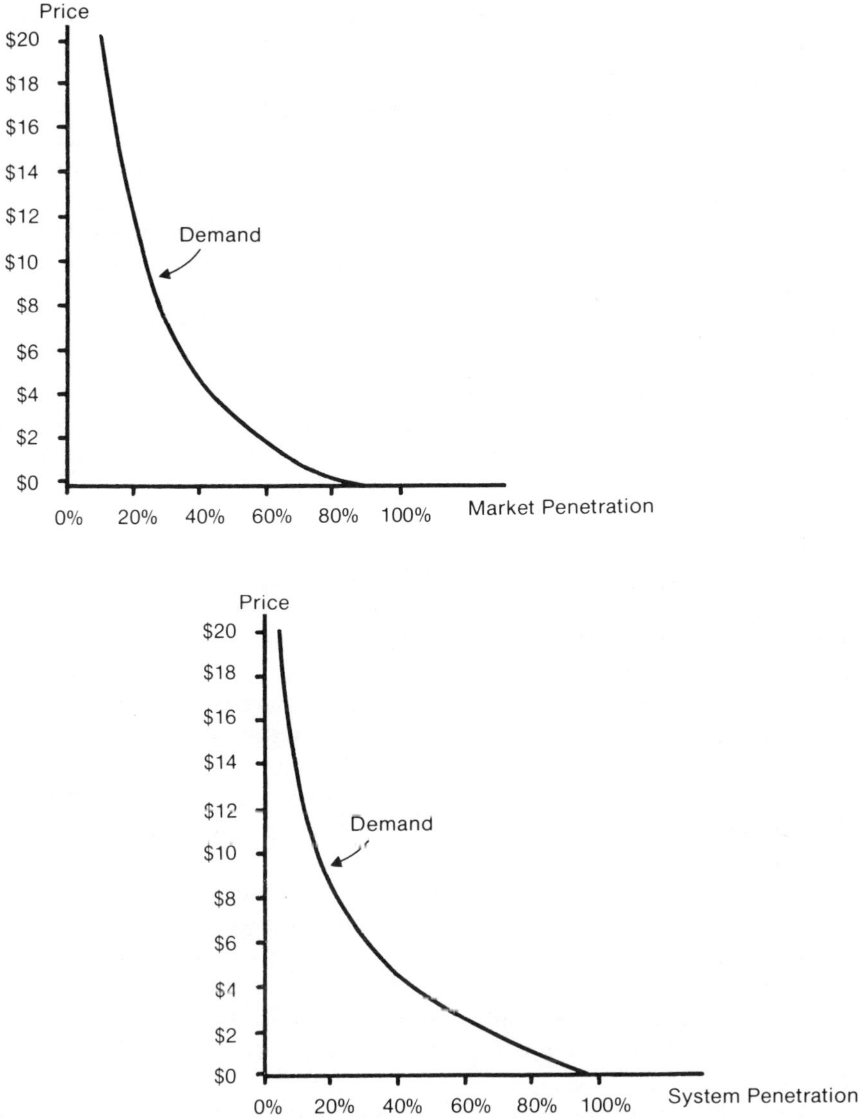

Note: The Philadelphia Market, 1979 dollars.

Figure 4-1. The Demand for Basic Cable (Top) and Pay Cable (Bottom)

system at 50-percent basic cable market penetration, a 5-percent decrease (increase) in basic-cable price can be expected to result in a 1.25-percent increase (decrease) in the number of pay-cable subscribers. The effect is more pronounced as pay-cable system penetration increases. Of course, if the system penetration of pay-cable were zero—no subscribers—changing basic-cable price would not be expected to affect pay cable.

For the cable operator, this complementary relationship means that pay-cable revenues can be increased by reducing basic-cable prices. The strategy can be compared to that of a restaurant owner who attracts customers by providing a lavish but inexpensive main course, and profits by selling expensive wine or desserts.

As another example, consider the strategy of a firm that produces cameras and also produces the only film that can be used in the cameras. The firm can offer cameras at very low prices to encourage sales and then make profits by selling the film at relatively high prices. Yet another example would be the case of an automotive manufacturer who might sell cars at less than cost if profits could be recouped through the sale of spare parts. In retailing parlance, a product sold at less than cost in this fashion is known as a *loss leader.* Supermarkets typically attract shoppers by putting a few items on sale, sometimes marked down below cost—the loss-leader pricing strategy.

In all these examples the pricing strategy requires that a firm has established a monopoly market for one good. The restaurateur has no immediate competitor when offering wine or dessert to customers. The camera firm provides the only film that can be used in its cameras. Similarly, the automotive manufacturer must have a monopoly on spare parts in order to recoup profits; otherwise, competition among spare-part suppliers would keep prices at levels too low to justify selling cars at less than cost. Intuitively it may seem that the loss-leader pricing strategy works to the consumer's disadvantage. As long as the firm earns no excess profits, however, consumers often benefit by paying lower average prices for goods and services.

For cable systems facing competition from subscription television and thus not having market monopoly, this pricing strategy may not be feasible. The great majority of cable systems, however, currently face little significant competition from subscription television. In these markets a pricing strategy that relies on pay-cable revenues to subsidize basic-cable prices appears to be efficient. Chapter 5 explores the nature of these pricing strategies in more detail.

Another relevant set of price elasticities defines the demand interrelationship among pay-cable services. In most cases these services are close substitutes, offering similar programming. The cross-price elasticities should be large and positive. The effect is difficult to quantify, however, because there is usually little variation among pay-cable prices when systems offer more than one pay-cable service. Because the services are close substitutes, market competition encourages price equality. The exceptions are pay-cable channels that offer as service some fraction of a full pay-cable channel at discounted prices.

A relationship that will become increasingly important as the industry matures defines the value of cable television as a medium for advertisement. The goal of the advertiser is to deliver a message to potential customers. Accordingly, the value of advertising time is determined largely by the number of viewers. The approximate 1980 value of a thirty-second slot of time in the broadcast media is $1 per 1,000 viewers. The more viewers a cable system can attract, the higher its expected advertising revenue. The relationship is complementary. As the price of cable increases, reducing the number of subscribers, the demand for advertising time on the cable system decreases. This effect will be moderated by the largely negative impact that advertising has on cable demand. Among the most important attributes of pay cable is that viewers are not obliged to endure advertisements. Viewer preferences are likely to vary, however. Some may be willing to accept a few commercials in exchange for lower prices. Varying the price of service and the advertising content of programming offers a means of product differentiation for pay-television networks.

Newspapers and magazines operate with a similar set of economic relationships. Subscription prices tend to be lower for publications that rely in part on advertising revenue. The combinations of advertising content and price offered by the print media range from free supermarket circulars filled with advertising to expensive economics books that contain no advertising. The variety of products distributed by the print media, with various combinations of advertising content and price, suggest a model of development for the pay-television industry.

Summary

Among the most important determinants of demand for basic cable are the number, type, and quality of signals carried by the system compared with those available over the air in the local market. To a large extent, basic cable still serves the market by providing improved reception of network-affiliated stations. Continued growth for the industry, especially in urban markets, will depend on service innovations such as increasing the number of satellite-distributed signals available to subscribers.

The importance of basic cable to low- and middle-income families may have been underestimated by the industry, which favors construction of systems in higher-income neighborhoods. The econometric model suggests that low-income neighborhoods exhibit high levels of demand.

No substitute services beside broadcast television measureably influence the demand for basic and pay cable. The current level of basic-cable prices, in the inelastic range, seem to reflect the effects of regulation. Raising basic-cable prices would increase the profits of the many systems that do not offer pay cable. For systems offering pay cable, a price hike for basic cable will tend to reduce pay-cable revenues. The relationship between basic and pay cable is complementary.

Appendix 4A
Estimating Demand

Data for a cross-section of 98 cable systems operating in cities throughout the country were taken from the *Broadcasting-Cable Yearbook: 1980* and verified against the independently collected and published data in the *Television Factbook* 1980. Additionally, a survey of 728 households was conducted in Philadelphia. The survey served both to supplement and to verify the national cross-section data. Aggregate socioeconomic characteristics for the cities in which the cable-television systems operated are from the *City and County Databook* (1977). A numeric description of the terrain for cities in the sample is from the *National Geologic Survey*.

The endogenous, or dependent, variables in the equations estimated from the national cross-section data are the market penetration (MP) of basic cable and the system penetration (SP) of pay cable. *Penetration* is the number of subscribers as a percentage of households in the market, a convenient form for the dependent variable because it permits comparisons among market areas of any size. A simple linear model, as in equation 4A.1, could be used to estimate demand equations for cable penetration.

$$\textit{Linear model: } P = XB + U \tag{4A.1}$$

where P is a vector of n observations on cable penetration, in decimal form.

X is a matrix of k explanatory variables, such as price or income, on n observations.

B is a vector of k coefficients.

U is a vector of disturbance terms.

One drawback to the linear model is that whereas actual values of P can range only from zero to one, representing market penetration from 0 to 100 percent, the linear model can produce such nonsensical forecasts for P as -0.15 or 1.20. A market-penetration figure of -0.15, or -15 percent, is meaningless. One solution to this problem is to estimate P with a logit, or log of the odds ratio, model.

The Logit Model

The logit employs a transformation of the dependent variable, *P,* which converts a percentage value to an odds ratio. Gambling odds are commonly expressed in odds-ratio form (equation 4A.2).

$$\textit{Odds ratio:}\ \frac{1}{1 - P} \tag{4A.2}$$

where *P* is cable penetration, expressed as a decimal.

As an example, consider a value of 0.25 for *P,* or that the market penetration for a given cable system is 25 percent. Inserting the value of *P* into the odds ratio, equation 4A.2, yields the fraction 1/3. That is, the odds that a household will subscribe are one to three: for every household that subscribes, three do not.

The goal of constraining the forecast value of *P* to range between 0 and 1 can be accomplished by estimating the natural logarithm (log) of the odds ratio, the logit model, specified in equation 4A.3. As the value of *P* approaches zero or one, the natural log of the ratio approaches negative or positive infinity. The form of the logit model is illustrated by figure 4A–1.

$$\textit{Logit model:}\ \text{Log}\left[\frac{P}{1 - P}\right] = XB + U \tag{4A.3}$$

where *P* is a vector of *n* observations. For the national cross-sectional data, *P* represents cable penetration. For the market-survey data, *P* represents the probability that an individual household will subscribe.

Log is the natural logarithm.

The vector of *k* coefficients in the model, *B,* measures the impact on the log of the odds ratio for a change in the value of an explanatory variable in *X,* an *n*-by-*k* matrix. When values in the matrix *X* are also in natural logarithm form, demand elasticities for the explanatory variables in the model, such as price and income, are equal to $(1 - P)B$.

That the demand elasticity depends on the penetration level is one of the intuitively appealing aspects of the logit model. It should be expected, for example, that a 5-percent change in price would affect demand differently for a system at 95-percent penetration than for one at 5-percent penetration.

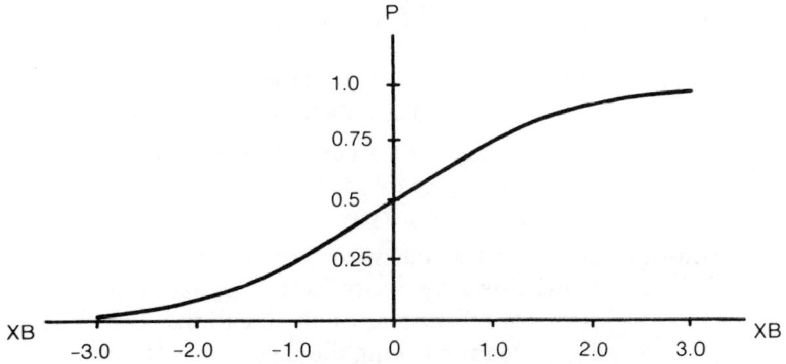

Figure 4A.1. Market Penetration (P) and Values for the Estimated Equation (XB)

A statistical method for testing the assumptions of the logit model is to compare the goodness of fit to the data for the logit model against alternative assumptions such as the linear model of equation 4A.1. The statistically superior model results in a lower sum of disturbances squared, as calculated with the untransformed dependent variable. That is, the superior model more closely fits the data. By this test, the logit model outperformed both linear and log-linear models of demand.

The Comparative-Service Variables

Among the most important variables that explain the level of demand for cable television, the comparative-service variables relate the quality of the cable service to the local broadcast market. The comparative-service variables are constructed to capture the effect of the number, type, and signal quality of available channels. Cable systems, it is assumed, deliver signals of good electronic quality.

The electronic quality of a broadcast signal declines with distance from the broadcast tower. Basic physical laws describe the attenuation of electromagnetic radiation, such as a broadcast signal, as being proportional to the square of distance. Therefore, the probability, $P_{ij}(R)$, that an individual viewer receives an adequate broadcast signal should also decline approximately as an inverse proportion to the square of the distance. In equation 4A.4 this assumption is combined with the FCC-measured value of $P_{ij}(R) = 0.5$ for a distance from the broadcast tower known as the B contour.

$$P_{ij}(R) = 0.5 + 0.5[d_{ij}]^2 \qquad (4A.4)$$

where $P_{ij}(R)$ is the probability that an individual viewer receives an adequate broadcast signal from station j of type i.

d_{ij} is the distance of a cable system to the B contour of station j, type i divided by the distance of the B contour to station j, type i.

B contour is a physical measurement of the distance at which viewers receive a satisfactory signal 90 percent of the time for 50 percent of the locations, provided by the FCC. Maps showing the B contours for broadcast stations operating in markets throughout the United States are available in the *Television Factbook* (1980).

The relationship defined by equation 4A.4 is illustrated in figure 4A–2. For example, if a viewer is in the market area of a cable system that is located very near a broadcast station, then d_{ij} equals 1 and $P_{ij}(R)$ also equals 1. In that event, and neglecting the effect of local interference from buildings or mountains, all viewers should receive a satisfactory signal.

If the cable system were located on the B contour of a broadcast station, the value of d_{ij} would be 0 and $P_{ij}(R)$ would equal 0.5, meaning that half the viewers in the cable system's market area would receive a satisfactory signal. The other half, who didn't receive a satisfactory signal, would

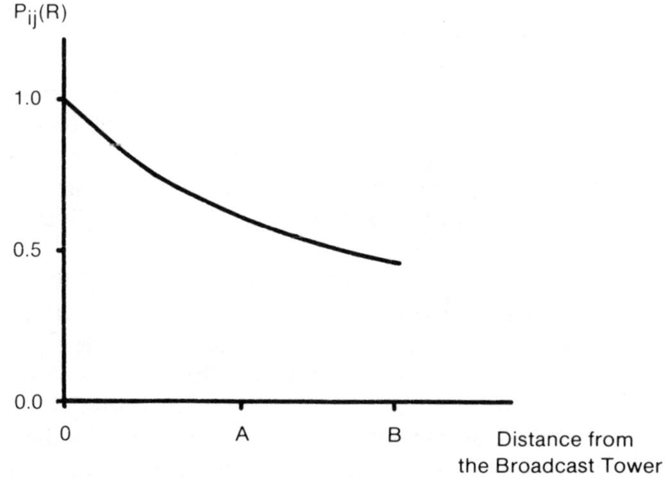

Figure 4A–2. The Probability of Good Television Reception as a Function of Distance to the Broadcast Tower

be potential cable-television subscribers. In the comparison of cable to broadcast service, equation 4A.5, the value of $P_{ij}(R)$ represents the percentage of households in the local market who can be expected to receive an adequate signal from each local broadcast station.

Comparative-Service Variable

$$X_i = \text{Log}\left[\frac{1 + S_i}{1 + \sum_{j=1}^{n} P_{ij}(R)}\right] \qquad (4A.5)$$

where X_i is the comparative service of the cable system for each station of type i for $j = 1, \ldots, 4$ types of signals: (1)(1) primary network, the nearest local affiliate for each of the three commercial networks; (2) duplicate network, additional network stations; (3) independent stations, those not affiliated with one of the three commercial networks, including Canadian stations and satellite networks; and (4) educational stations.

S_i is the number of signals of type i carried on the cable system.

$P_{ij}(R)$ is for station j, where $j = 1, \ldots, n$ stations in the local market of type i, the probability that the median observer in the city in which the cable system is located receives satisfactory over-the-air reception at least 90 percent of the time (see equation 4A.4). A probability of zero is assigned for satellite networks.

The logarithmic-ratio form of the comparative-service variable is suggested by the economic law of diminishing marginal utility. Given this form, each additional signal introduced by the cable system will have a larger impact on demand in markets where there are few existing available signals than in markets where many signals are already available.

Another factor that may affect the quality of local broadcast signals and that has been included in previously published studies (Park 1971) is that UHF signals are more succeptible to local interference than are VHF signals. In general, the A and B contour for UHF stations will be closer to the station's broadcast tower than for VHF stations. This information is included in equations 4A.4 and 4A.5. Adding another parameter to the comparative-service variable that further discounts UHF signal quality, however, did not measurably improve the model.

Demand Equations from the National Cross-Section Data

Table 4A–1 describes equations that use the comparative-service variable and other relevant determinants of demand to model the expected market penetration of basic cable and system penetration of pay cable. Several hundred specifications of the two equations were estimated to determine the final form of each of the exogenous or independent variables described in the table. In general, exogenous variables were deleted from the equation

Table 4A–1
The Demand for Basic Cable and Pay Cable as Estimated from a National Cross-Section
(1979 dollars)

Dependent Variables

for basic cable:

$$\text{Log}\left[\frac{P_b}{1 - P_b}\right]$$

where P_b = the number of households that subscribe to the basic system divided by the number of households passed by cable

for pay cable (HBO):

$$\text{Log}\left[\frac{P_p}{1 - P_p}\right]$$

where P_p = the number of households that subscribe to pay cable (HBO) divided by the number of households that subscribe to the basic system

Independent Variables	*Basic Cable (s.e.)*[a]	*Pay Television (s.e.)*[a]
Constant	7.432	−18.178
Log[b] Monthly subscription price for the basic-cable system + 1)	−1.031 (0.37)[a]	
Log Monthly subscription price for Home Box Office + 1)		−1.717 (0.38)[a]
X_1: Comparative-service variable for network signals (see equation 4A.5)	1.390 (0.29)	−0.901 (0.38)
X_2: Comparative-service variable for duplicate network signals (see equation 4A.5)	0.237 (0.09)	
X_3: Comparative-service variable for independent signals (see equation 4A.5)	0.410 (0.10)	
The number of cable channels dedicated to public-access programming	0.061 (0.08)	
The number of cable channels dedicated to locally originated programming (but not public access)	0.066 (0.08)	

Table 4A-*1 continued*

Independent Variables	*Basic Cable (s.e.)*[a]	*Pay Television (s.e.)*[a]
Log (number of months since the system franchise was granted)	0.239 (0.06)	
(Geological Survey Scale ranging from 1 to 20 describing the local terrain, 1 being flat, 20 being mountainous) squared	0.015 (0.01)	
Log (per capita income in 1974 dollars)	−1.232 (0.41)	2.481 (0.974)
Log (index of Home Equipment[c])	0.197 (0.09)	
Log [100 (percentage of population having completed high school)]	0.471 (0.33)	
1 if the system is located in the Mountain West[d] 0 otherwise	0.221 (0.06)	
1 if the system is located in New England[e] 0 otherwise	0.324 (0.33)	
1 if the system is located in the South Atlantic[f] 0 otherwise	0.431 (0.14)	
1 if the system is located in the West North Central[g] 0 otherwise	−0.455 (0.19)	
1 if the system offers pay cable 0 otherwise	0.003 (.108)	

Equation Statistics

	Basic Cable	*Pay Television*
R^2	0.692	0.25
$\bar{R}^2$	0.632	0.21
σ	0.453	1.173
F	11.530	6.840
N	98	86

Note: For a discussion of the logit model as applied to cable television, see Charles River Associates (1973).

[a]s.e. = standard error of the estimate, included in parentheses.

[b]Log indicates that values in parentheses are transformed by the natural logarithm.

[c]The Index of Home Equipment is derived by adding together the percentages of occupied units with air conditioning, home food freezer, telephone, and one or more automobiles. It reflects consumer spending patterns.

[d]The Mountain West includes Arizona, Colorado, Idaho, Montana, Nevada, New Mexico, and Wyoming.

[e]The Northeast includes Connecticut, Maine, Massachusetts, New Hampshire, Rhode Island, and Vermont.

[f]The South Atlantic includes Delaware, Florida, Georgia, Maryland, North Carolina, South Carolina, and Virginia.

[g]The West North Central includes Iowa, Kansas, Minnesota, Nebraska, North Dakota, and South Dakota.

unless the probability of a type-one error—wrongly concluding that the variable is different from zero when it is not—was about 10 percent or less. A dummy variable, set to one for systems offering pay television, was included as evidence that the presence of pay television at normal prices does not affect basic-cable demand. The coefficients described in table 4A–1 are the measures of demand discussed throughout chapter 4.

Demand Equations from the Philadelphia Market Survey

The market survey of Philadelphia presented households with a series of binary choices: to subscribe or not subscribe. In each case the utility of the two choices, subscribing or not subscribing, is given by equations 4A.6 and 4A.7.

$$Y_1 = X_1B_1 + U_1 \tag{4A.6}$$

$$Y_2 = X_2B_2 + U_2 \tag{4A.7}$$

where Y_i is the utility of each choice.

i is 1 when the choice is to subscribe.

i is 2 when the choice is not to subscribe.

Signifying a choice to subscribe as $Y = 1$, then

$$Y = 1 \text{ if } Y_1 > Y_2 \tag{4A.8}$$

$$Y = 0 \text{ if } Y_2 > Y_1 \tag{4A.9}$$

If the error terms U_1 and U_2 are independent standard Weibull (extreme value) distributed, the ex ante probability that $Y = 1$ is the cumulative distribution described in equation 4A.10.

$$P = \frac{1}{1 + \exp(-XB)} \tag{4A.10}$$

where P is $P_r[Y = 1]$.

A few manipulations of equation 4A.10 yield the logit model, equation 4A.3, which is the form of the estimated equations in table 4A–2 summarizing the results of a market survey of Philadelphia. The survey consisted of 728 telephone interviews conducted during 1981 and 1982. Equations

Table 4A–2
The Demand for Basic Cable and Pay Television as Estimated from a Market Survey of Philadelphia
(1979 dollars)

Dependent variables

(for basic cable) $\text{Log}\left[\dfrac{P_b}{1 - P_b}\right]$ (for pay-cable services) $\text{Log}\left[\dfrac{P_i}{1 - P_i}\right]$

where P_b = the number of households that subscribe to the basic system divided by the number of households passed by cable.

P_i = the number of households that subscribe to service type i divided by the number of households that subscribe to the basic system.

i = for i = 1, 4 types of service: (1) Showtime, a pay-TV entertainment service; (2) HBO, a pay-TV entertainment service: (3) PRISM, a pay-TV entertainment service with an emphasis on area sports events; (4) Economy Movies, a pay-TV entertainment service with a limited selection of recent movies.

	Independent Variables		
Dependent Variables	*Constant*	*Log (Monthly Subscription Fee + 1)*	*Log (Household Income)*
Basic cable (σ) (t-stat)	−5.070	−1.035	0.86987
	(2.04)	(.216)	(.204)
	(2.49)	(4.80)	(4.35)
Showtime (σ) (t-stat)	2.723	−1.632	
	(.608)	(.273)	
	(6.09)	(5.99)	
HBO (σ) (t-stat)	3.946	−1.667	
	(.648)	(.266)	
	(6.09)	(6.30)	
PRISM (σ) (t-stat)	−1.876	−1.321	0.51656
	(2.17)	(.251)	(0.215)
	(.86)	(5.27)	(2.40)
Economy Movies (σ) (t-stat)	0.183	−0.771	
	(.335)	(.254)	
	(.547)	(3.18)	

were estimated using the computer package Crawtran developed by Robert B. Avery of Carnegie-Mellon University. Measured price elasticities for basic- and pay-cable, HBO, are nearly identical for both the national cross-section (table 4A–1) and the market survey (table 4A–2).

Two significant measurement problems were encountered with the market survey that, without adjustment, would produce an upwardly biased estimate of the level of demand. First, there was a problem of missing observations. One-third of the respondents expressed a desire not to answer the survey. In many cases a lack of enthusiasm for the survey was related to a lack of interest in cable television. This relationship was not difficult to identify; often the information was volunteered. In order to minimize the problem of missing observations, all respondents who declined to answer the complete survey were asked if they had any interest in subscribing to cable television. Almost all were willing to answer this single question with at least a yes or a no. Respondents who expressed interest were assumed to have the same demand for cable as households answering the entire survey. It was also assumed that respondents who expressed no interest in cable would be unwilling to subscribe unless service were offered free of cost.

The second source of market-survey bias was that households overestimated their own willingness to subscribe to cable. This tendency was suspected by the students who conducted the survey. It was verified by comparing the results of the market survey to the actual market penetration achieved by cable in a part of the city where service is available. The overestimate was about the same for each cable service. In order to correct for this tendency, the constant term for each of the equations derived from the market survey was reduced by a factor of 1.33. With this adjustment and the adjustment for missing data, the market-survey equations are consistent both with the equations derived from the national cross-sectional data and with the observed behavior of households in the city who have access to cable television.

A related problem complicates the income data in the market survey. The average household income as reported in the survey is nearly twice as high as would be expected from a random sample of the city. Respondents to survey interviews often overstate their actual incomes, presumably to impress the individual who is conducting the survey. Contributing to this problem, some of the students who conducted the survey interviewed friends and family instead of randomly selecting respondents as they had been instructed. Since the students were themselves from a higher-than-average income group, so were their family and friends. The survey sample, therefore, contains responses from more high-income households than would a random sample.

Among these high-income households, the demand for basic cable is positively correlated with income. Accordingly, the estimated coefficient for income in the basic-cable equation of table 4A–2 is positive. Among the households included in the cross-section sample, where incomes averaged in the middle- to lower-middle range, basic-cable demand was negatively correlated with income. The cross-section sample also includes weak statistical evidence that, for high-income families, the correlation between basic-cable demand and income is positive.

The combined evidence for these different income samples seems to be that the correlation between basic-cable demand and income is dependent on the level of income, being negative for low to middle income levels and positive for high income levels. Individually, neither sample contained enough information to verify this relationship.

Statistical Issues in Model Building

Identifying interrelationships among variables is a necessary step in the estimation of reliable econometric models. In general, good estimates of model coefficients—B in equation 4A.3—require that the disturbance terms—U in equation 4A.3—display the following properties: zero mean, constant variance, and pairwise uncorrelated. That is,

$$E(U) = 0 \tag{4A.11}$$

$$E(UU') = \sigma^2 I \tag{4A.12}$$

where σ^2 is the variance and I is the identity matrix.

The estimated equations described in chapters 3 and 4 were tested statistically for symptoms of autocorrelation or heteroskedasticity, two often encountered violations of the requirements described by equations 4A.11 and 4A.12.

Autocorrelation

A violation of the assumption that the off-diagonal elements of $E(UU')$ are nonzero, autocorrelation means that the disturbance terms among observations are interrelated. A likely cause of autocorrelation is a misspecified model—for example, the use of a linear model when the actual relationship more nearly resembles the logit. The relationship between the level of income and the effect of income on basic-cable demand would have resulted in autocorrelation if the relationship had been more pronounced in either sample. Autocorrelation may also be caused if important variables are excluded from the model. For that reason, many potentially important variables were tested for their relevance.

When autocorrelation is present in an equation, the statistical tests for the significance of estimated coefficients are invalid. Also, the standard error of the equation will be larger than necessary. That is, more uncertainty will appear to exist than actually does exist.

The Durbin-Watson statistic was used to test for the presence of autocorrelation. Because it tests only for autocorrelation between successive disturbance terms, the Durbin-Watson statistic was calculated with the cross-sectional data arranged in the order of magnitude for a particular variable in the model. The test was then applied successively, ordering the data for each variable in each equation. Where the value of the Durbin-Watson statistic revealed the presence of autocorrelation, variables were respecified. The linear form of the penetration variable, for example, demonstrated positive first-order autocorrelation, overestimating penetration at low levels. Respecifying penetration in the logit form eliminated the autocorrelation.

Heteroskedasticity

A violation of the assumption that the diagonal elements of $E(UU')$ are of the same size, heteroskedasticity means that the size of the disturbance terms are correlated with the size of one of the variables. As with autocorrelation, the usual tests for the statistical significance of estimated coefficients are invalid, and the standard error of the equation is larger than necessary in the presence of heteroskedasticity. To test for heteroskedasticity, the absolute values of the disturbance terms were regressed against each of the variables in the model, a test suggested by Glejser (1969). No significant heteroskedasticity appeared to exist in the demand model. The correction for heteroskedasticity discovered in the cost equations is described in appendix 3A.

Multicollinearity

Another common statistical problem, multicollinearity means that the independent, or exogenous, variables are interrelated and the separate influences of each variable are difficult to estimate. The estimated coefficients are also both imprecise and unstable. Although no statistical tests measure the severity of the multicollinearity problem, the low level of intercorrelation among the independent variables in the model and the stability of the estimated coefficients for several hundred model specifications suggest that multicollinearity is not a problem.

5 Pricing Cable Services

Two economic characteristics of cable television complicate the problem of pricing cable services. One is a cost characteristic. The results presented in chapter 3 indicate that cable-television systems are natural geographic monopolies, exhibiting declining average total costs. Consequently, the pricing strategies applicable to cable systems differ somewhat from those attributed to purely competitive firms. The first part of this chapter considers the problem of pricing for a natural monopoly.

The other economic characteristic of cable television that complicates the pricing problem is the presence of demand-interdependent cable services, described in chapter 4. Changing the price of basic cable affects the demand for pay cable. The cable industry has responded to this relationship with a variety of pricing strategies. One company, Cablevision, won the franchise award to build a system in Boston by offering low prices for basic cable; then, the following year, Cablevision proposed to build a system in San Francisco that would have high basic-cable prices but free pay cable. The two strategies appear to be a contradiction. Was one a mistake?

A method for analyzing these subtle pricing problems originated in an article by Frank Ramsey (1927). The method, known as Ramsey pricing, includes the proposition that consumers who are insensitive to price changes should pay higher prices than those who are not. Both business firms and society can advantageously exploit this proposition. For example, this chapter will show how the two apparently contradictory pricing strategies of Cablevision both represent efficient responses to a particular set of market conditions. The chapter also includes several empirical examples that illustrate the logic of cable-television pricing decisions.

Pricing Strategies for a Natural Monopoly

The fundamental logic of cable-television pricing, as opposed to the textbook logic of purely competitive markets, depends on the natural monopoly status of cable systems. Figure 5–1 illustrates three pricing strategies for a natural monopoly that offers one service such as basic-cable television. They are: profit maximization, social-welfare maximization, and con-

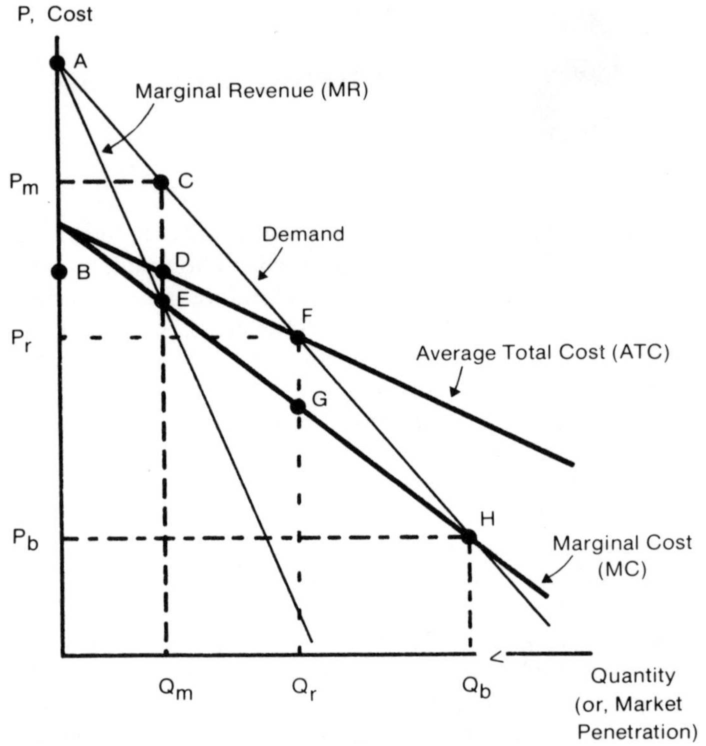

Notes:
P_m the monopoly price ($MR = MC$).
P_r the Ramsey price ($P = ATC$).
P_b the "best" price ($P = MC$).

Figure 5-1. Three Pricing Strategies for a Natural Monopoly

strained social-welfare maximization. The latter is commonly referred to as Ramsey pricing. A mathematical formulation of these three strategies, written to illustrate the method of Ramsey pricing, appears in equation 5.1. A formal derivation of the equation, expanded to include the case of a firm offering more than one service, is included in appendix 5A.

The Ramsey number (λ) *for three pricing strategies:*

$$\left[\frac{P - MC}{P}\right]\eta = -\lambda \tag{5.1}$$

where P is the price of the service.

MC is the marginal cost of the service, defined to be the change in total cost (TC) divided by the associated change in output (Q) of service, or $\Delta TC/\Delta Q$.

η is the price elasticity of demand; as defined in chapter 4, the percentage change in the quantity (Q) of the service demanded divided by the associated percentage change in price (P), or $\%\Delta Q/\%\Delta P$.

λ is the Ramsey number, taking values from zero to one depending on the pricing strategy: $\lambda = 1$ for profit maximization; $\lambda = 0$ for social-welfare maximization; $0 \leq \lambda \leq 1$ for constrained social-welfare maximization, Ramsey pricing.

Profit Maximization

A firm seeking to obtain the highest short-run profit maximizes the difference between total revenue and total cost. In terms of equation 5.1, the profit-maximizing monopolist would set a price so that the Ramsey number equals one. With some rearranging, equation 5.1 can be written as follows when the value of the Ramsey number is one.

$$MC = P\left[1 + \frac{1}{\eta}\right] \text{ or } MC = MR \tag{5.2}$$

for $\lambda = 1$, where MR is the marginal revenue for a one-unit increase in output; defined as the change in total revenue (TR) divided by the associated change in output (Q), or $\Delta TR/\Delta Q$.

Equation 5.2, which equates marginal cost with marginal revenue, represents the traditional economic rule of short-run profit maximization. For the monopolist characterized by figure 5–1, marginal cost equals marginal revenue at point E. To maximize profits, the monopolist should choose to produce at point Q_m. At that output level the monopolist can charge price P_m, represented by point C on the demand line.

With this strategy the monopolist earns a profit, in excess of normal profits, represented by the area in the rectangle formed by points P_m, B, C, and D. The area is equal to total revenue less total cost. Normal profits, included in the economic definition of costs, refers to the level of profits a firm would accept in competitive equilibrium. In chapter 3, normal profits for a cable-television firm were estimated to be about 8.2 percent, adjusted for inflation. To achieve higher profits, the monopolist exercises market power by restricting output and raising prices above competitive levels.

Perhaps the most notorious recent example of this effect was the increase in oil prices imposed by the Organization of Petroleum Exporting Countries (OPEC) during the 1970s. Acting in concert, the OPEC countries gained market power over oil.

At the end of the nineteenth century Bell Telephone established market power over the telephone service. Protected from competition by patent, and not yet regulated, Bell set prices at a level that generated high profits. The response of government to the potential for monopoly pricing has been the regulation of prices for local telephone service. In several states the regulation of basic-cable prices has been similarly justified.

Social-Welfare Maximization

The objection to an unregulated monopolist is that prices will be higher and output lower than expected from firms in competition. In general the competitive market is the benchmark for pricing efficiency. Competitive prices maximize equation 5.3.

$$\textit{The social-welfare criterion: } W(P) = S(P) + \Pi(P) \qquad (5.3)$$

where $W(P)$ is aggregate social welfare at price P.

$S(P)$ is consumer's surplus, given by the area under the market-demand line and above the market price.

$\Pi(P)$ is producer's surplus, given by the area above the supply line and below the market price P, representing profits and economic rent.

The notion of consumer's surplus in equation 5.3 plays a major role in the following pages; it represents the value of service in excess of the market price. In figure 5-1 consumer's surplus is the area in the triangle formed by points A, C, and P_m when the monopolist's price prevails. It is the sum over all consumers of the difference between the price a consumer is willing to pay for service and the market price. The demand line schedules the maximum price consumers are willing to pay.

An example sometimes used to illustrate the concept of consumer's surplus is the comparative value of water and diamonds. Water is less expensive but, of course, more valuable to society. The difference in value is not reflected by the market price. In the marketplace, water is a bargain because of its abundance. The value of water to society in excess of the market price is approximately equal to the area below the demand line and above the market price—the area of consumer's surplus. There is a great deal of consumer's surplus associated with water, and relatively little associated with diamonds. A salesman who could charge exactly the maximum price that each consumer was willing to pay could convert any consumer's surplus into potential revenue. Thus it can be useful to think of consumer's surplus as potential revenue. The potential revenue of water is much greater than the potential revenue of diamonds; therefore, water is more valuable.

Actual, as opposed to potential, revenues contribute to producer's surplus, the profits and rent earned by firms supplying the services. The sum of producer's and consumer's surplus is the social-welfare criterion, as defined in equation 5.2. Firms in competitive markets automatically maximize the social-welfare criterion by choosing to produce at levels described by equation 5.4, where the Ramsey number equals zero.

$$\textit{Competitive pricing:} \quad \left[\frac{P - MC}{P}\right]\eta = 0 \qquad \text{or} \qquad P = MC \qquad (5.4)$$

for $\lambda = 0$.

When competitive markets are in equilibrium, firms will earn just normal profits and provide service to everyone who is willing to pay the marginal cost of production. For a natural monopoly, however, marginal cost is less than average total cost. Consequently, a natural monopoly operates at a loss when price is equal to marginal cost. For example, the natural monopoly described in figure 5–1 would find that price equals marginal cost at point *H,* which corresponds to output level Q_b and price P_b. Although the firm would maximize social welfare at price P_b, each unit of output would be sold at less than average cost of production. The firm would lose money on each sale.

Of course, no firm would voluntarily continue operating at a loss for a long time. A natural monopoly regulated to operate at the level of price equal to marginal cost must, therefore, be subsidized if operations are to continue. For example, public-transportation systems are often subsidized in order to keep prices low. If the firm is not subsidized and is to continue operations, prices must be high enough to meet costs. The natural monopoly must then deviate from the price-equals-marginal-cost rule. The best price, which maximizes the social-welfare criterion, is not feasible.

Ramsey Pricing: The Constrained Social-Welfare Maximization

Given the constraint that the firm should be allowed to earn at least a normal profit, the price that maximizes the social-welfare criterion for a single-product firm is equal to average total cost (ATC). In figure 5–1 the price P_r associated with output level Q_r allows the firm to break even, earning just a normal profit. For a single-product natural monopoly, the break-even price is called the second-best solution to the social-welfare problem because it results in a lower value for the social-welfare criterion than the so-called best price, determined by the price-equals-marginal-cost rule of equation 5.4. The break-even price is also the Ramsey price. It is the price that results in the lowest value for the Ramsey number, described in equation 5.1, which fulfills the constraint of allowing the firm to earn a normal profit. In general, the value of the social-welfare criterion increases as the Ramsey number decreases.

A regulator with economic goals, but who does not want to subsidize the operation of a firm, should choose the break-even Ramsey price. In fact, regulators typically do select prices that are intended to allow a normal profit for the firm. This regulatory approach is applied to local telephone service, many public utilities, and some cable-television systems subject to rate-of-return regulation. The prices regulators set for these services are designed to provide a targeted normal profit for the regulated firm.

Of course, regulation also has its costs, both the direct costs of operating the regulatory mechanism and the indirect costs of imperfect regulation. A diminished incentive to innovate seems to be one of the indirect costs. Critics of regulation assert that the local telephone exchange has been slow to develop technologically because government regulation removes the incentive to innovate. If this is true, the indirect cost to society is less efficient telephone service and prices that are higher than necessary.

A problem associated with rate-of-return regulation is that the regulated firm may have an incentive to overinvest in capital and corporate amenities. The firm can absorb these costs and increase dollar profits if the regulatory authority raises prices to achieve a targeted rate-of-return. Again, the indirect cost to society is higher than necessary prices. Regulatory authorities have attempted to avoid this indirect cost of regulation by carefully reviewing investments to be included as the basis for determining rates-of-return. Of course, disputes have arisen over what costs should be included; often the result is litigation, which increases the direct cost of regulation.

The question of whether or not to regulate an industry is often approached as one of determining which is less expensive for society, the cost of regulation or that of monopoly pricing. It is often a difficult question to resolve. A different approach to the problem has been taken by municipal authorities who award cable-television franchises by competitive bidding. The practice is intended to provide monopoly services at competitive prices without extensive government regulation and its associated costs. Firms bidding competitively may be expected to choose the lowest possible price that allows normal profits. Will the franchise-winning firm, however, maintain the competitive price for the duration of the franchise agreement? Often the firm specifies future price changes as part of the franchise proposal. This apparent solution has its drawbacks. Given the uncertainties of inflation and service innovation, it is difficult to establish what constitutes competitive prices for a franchise agreement that may extend fifteen years into the future.

When entry into the market is free and exit is costless, even natural monopolies may be expected to maintain competitive prices. During the early years of cable franchising, a few municipalities made the mistake of awarding exclusive franchises with no time limitation. These contracts made entry illegal. More recently, franchise awards have not been exclusive, allowing for the possibility of competitive entry. For cable systems, however, both entry and exit from the market can be very costly. The initial investment is very high, and the equipment used by cable systems is specific to a market area so it cannot easily be sold.

An analysis of this problem associated with the so-called Chicago school of industrial organization and the writings of Joseph Schumpeter suggests that monopolistic pricing will stimulate innovations designed to erode any market barrier. Also, the high prices will shift consumption into other services. For example, high oil prices during the 1970s stimulated attention to alternative fuel sources and a few years of reduced demand. Of course, this process requires time and is not without cost to society.

Pricing for Multiple-Service Cable Systems

Many cable systems offer more than one service, basic cable and one or more pay-cable channels. For a multi-service firm, determining the competitive, Ramsey prices can be complicated. Almarin Phillips notes that a wide variety of prices may result from regulations that are not expected from a marginal-cost perspective (1980). For example, at least one of the products or services of a firm may be priced above the price a monopolist would be expected to charge, or below marginal cost.

Efficient prices, which maximize the social-welfare criterion subject to the constraint of a normal profit for the firm, can be determined by expanding the Ramsey-pricing rule of equation 5.1 to include the price, price elasticities, and marginal costs for each of the products or services sold by the firm. A mathematical derivation of the rule is included in appendix 5A.

In general, prices should be set so that the percentage deviation from marginal cost times the elasticity of demand is constant among all products or services. The effect of price changes on income distribution, the demand for other goods or services in the economy, and the effect on events external to the market should also be considered. In the presence of declining average total costs, Ramsey prices are achieved when the Ramsey number is the same for all services and takes the lowest value that allows a normal profit. These Ramsey prices are efficient from the perspectives both of society and of a competitive firm. For example, a multiservice cable-television firm bidding against other firms to win a cable franchise should adopt a Ramsey-pricing strategy if the municipality awarding the franchise has economic goals, as described in equation 5.3. A cable firm already operating as a monopoly in a market, but fearing competitive entry from other firms offering related services, may also find that Ramsey prices represent the best long-run pricing strategy for discouraging competitive entry. Ramsey prices are competitive in markets characterized by declining average total costs, when the firm is a natural monopoly.

For a multiservice firm, the least complicated pricing environment exists when services are demand independent—that is, when a change in the price of one service does not affect the demand for another. If so, cross-price elasticities are zero, and the following equation applies for a two-service firm.

Ramsey pricing for two demand-dependent services

$$\left[\frac{P_1 - MC_1}{P_1}\right]\eta_1 = \left[\frac{P_2 - MC_2}{P_2}\right]\eta_2 = -\lambda \tag{5.5}$$

For the lowest value of λ that allows the firm to earn a normal profit, where P_1, MC_1, and η_1 refer respectively to the price, marginal cost, and elasticity of demand for service 1; and P_2, MC_2, and MC_2, and η_2 refer respectively to the price, marginal cost, and elasticity of demand for service 2.

The value in the brackets of equation 5.5, price minus marginal cost divided by price, represents the percentage by which price is marked up over marginal cost. It is useful to think of Ramsey pricing as the optimal markup. According to equation 5.5, the service with the more inelastic demand—a lower absolute value of η—must have the higher markup. As illustrated in figure 5–2, raising the price of the demand-inelastic service reduces consumer's surplus by less than an equivalent increase in the price of the demand-elastic service. Notice also in figure 5–2 that fewer consumers are excluded from the market when the price is raised for the demand-inelastic service than when price is raised for the demand-elastic service. Ramsey pricing is designed to keep output at the highest level that allows the firm to earn a normal profit.

According to equation 5.5, when the demand elasticities and marginal costs of a service are the same, prices should be the same. This conclusion is illustrated in figure 5–3 along with some of the details of what Ramsey accomplishes.

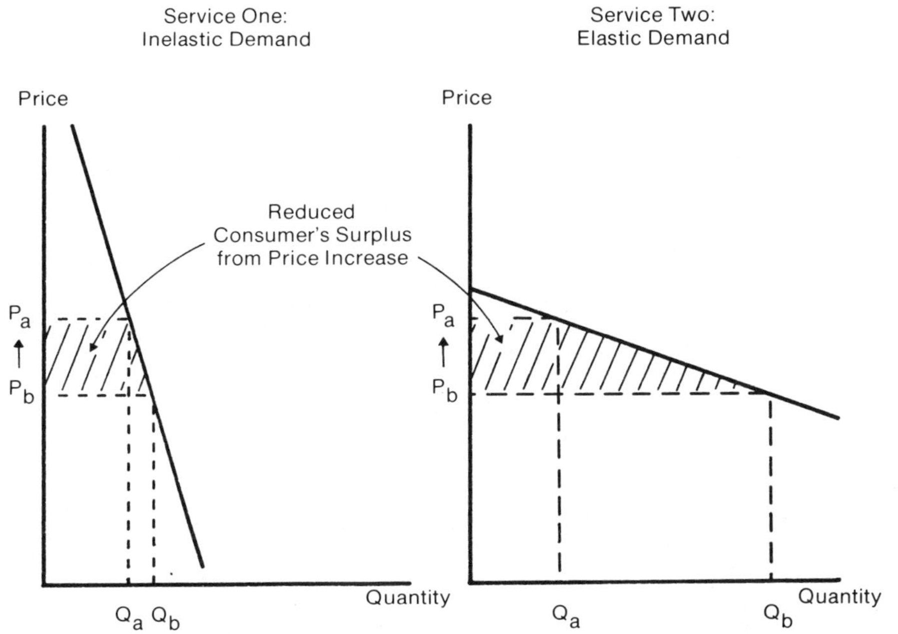

Notes:
P_b and P_a are the price of service before (P_b) and after (P_a) the price increase.
Q_b and Q_a are the quantity of subscribers served before (Q_b) and after (Q_a) the price increase.

Figure 5-2. The Change in Consumer's Surplus Resulting from an Increase in Price

A complication of pricing for a multiple-service firm is that many different price combinations may yield a normal profit, allowing the firm to break even. Recall that Cablevision has used both the strategy of low-priced basic-cable with high-priced pay cable, and the reverse strategy of high-priced basic cable with low-priced pay cable in bidding for cable-television franchises. For the two services offered by the hypothetical firm described in figure 5-3, the price combinations that allow the firm to break even are shown by a profit isocontour labeled $\lambda = 0$; for these price combinations, the firm earns zero excess profits. An isoprofit contour could be drawn for any feasible profit level. Maximum profits are achieved at point *A,* where prices are set so that marginal revenue equals marginal cost.

Once profits are identified, the next aspect of the pricing problem concerns social welfare. The level of producer's surplus that contributes to social welfare is fixed for the natural monopoly by the requirement of zero excess profits. Only the consumer's surplus component of social welfare is variable. The Ramsey-pricing problem may be solved, then, by finding the

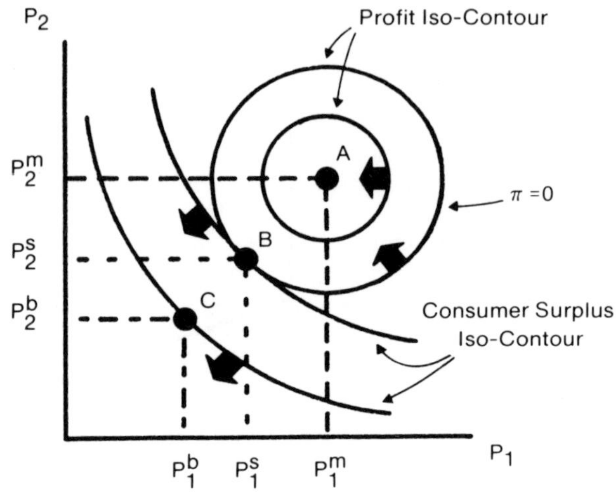

Notes:

P_1^m is the profit-maximizing price for service 1.
P_2^m is the profit-maximizing price for service 2.
P_1^b is the social-welfare-maximizing price for service 1—the "best" price.
P_2^b is the social-welfare-maximizing price for service 2—the "best" price.
P_1^r is the Ramsey, or constrained social-welfare-maximizing price for service 1.
P_2^r is the Ramsey, or constrained social-welfare-maximizing price for service 2.
A is the profit-maximizing set of prices.
B is the Ramsey, or constrained social-welfare-maximizing set of prices.
C is the social-welfare-maximizing set of prices.

Figure 5–3. Ramsey Pricing for Two Demand-Independent Services

prices that maximize consumer's surplus and yield a normal profit for the firm.

As with profits, a given level of consumer's surplus may be achieved by many different price combinations. Two consumer's surplus isocontours are illustrated in figure 5–3, each representing a set of prices that yields the same level of consumer's surplus. As prices decline, consumer's surplus increases; the isocontour closest to the origin in the figure represents the highest level of consumer's surplus. The tangency of the consumer's surplus isocontour closest to the origin and the zero-excess-profits isocontour identifies the Ramsey prices. At the point of tangency, consumer's surplus is the highest for any set of prices that allows the firm to earn a normal profit. The tangency occurs at point *B* in figure 5–3. Because the offered services have identical marginal costs and demand elasticities, the Ramsey prices for the two services are also identical.

The dotted line connecting point *A* to point *C* in the figure represents all the tangencies between consumer's surplus and profit isocontours, solutions to the Ramsey-pricing rule of equation 5.5, for $0 \leq \lambda \leq 1$. At point *A,* where $\lambda = 1$, the firm earns a monopoly profit. At point *C,* where $\lambda = 0$, prices are equal to marginal costs but the firm fails to break even, so the solution is not feasible. The Ramsey-pricing problem for a natural monopoly may be solved by finding the prices associated with the lowest value of λ that allows the firm to earn a normal profit.

The same approach is applicable when the services offered by the cable system are demand interdependent—that is, when a price change for one service affects the demand for another; the Ramsey pricing rule must be expanded to capture the interdependence, however, as in the following equation.

Ramsey pricing for two demand-interdependent services:

$$\left[\frac{P_1 - MC_1}{P_1}\right] \eta_{1,1} + \left[\frac{P_2 - MC_2}{P_1}\right] \frac{x_2}{x_1} \eta_{2,1} =$$

$$\left[\frac{P_2 - MC_2}{P_2}\right] \eta_{2,2} + \left[\frac{P_1 - MC_1}{P_2}\right] \frac{x_1}{x_2} \eta_{1,2} = -\lambda \qquad (5.6)$$

for $0 \leq \lambda \leq 1$ where $\eta_{1,1}$ and $\eta_{2,2}$ represent the own-price elasticities for services one and two, $\eta_{1,2}$ represents the cross-price elasticity of demand for service 2 with respect to a change in the price of service 1, and $\eta_{2,1}$ represents the cross-price elasticity of demand for service 1 with respect to the price of service 2.

The upshot of equation 5.6 is that in setting the price of any service, the demand impact on all services must be considered. For a cable system offering basic cable and pay cable but not having the technical constraint that only basic-cable subscribers may subscribe to pay cable, the services would be substitutes and the cross-price elasticities would be positive. The cross-price elasticities among pay-cable services are also expected to be positive.

Figure 5–4 illustrates the pricing problem for a firm offering two substitute services characterized by identical marginal costs, own- and cross-price elasticities. Notice that the consumer's surplus isocontours are more convex than in the demand-independent case of figure 5–3, and the profit isocontours are elliptical instead of circular. The shapes of these contours indicate that if either profits or consumer's surplus is to be held at a constant level, a price reduction for one service must be offset by a larger price increase for the other service than is needed when services are independent. The potential for pricing inefficiency is most severe when services are sub-

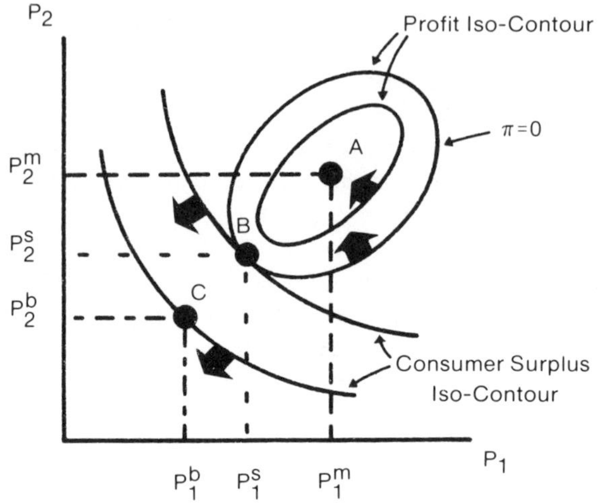

Notes:

P_1^m	is the profit-maximizing price for service 1.
P_2^m	is the profit-maximizing price for service 2.
P_1^b	is the social-welfare-maximizing price for service 1—the "best" price.
P_2^b	is the social-welfare-maximizing price for service 2—the "best" price.
P_1^r	is the Ramsey, or constrained social-welfare-maximizing price for service 1.
P_2^r	is the Ramsey, or constrained social-welfare-maximizing price for service 2.
A	is the profit-maximizing set of prices.
B	is the Ramsey, or constrained social-welfare-maximizing set of prices.
C	is the social-welfare-maximizing set of prices.

Figure 5-4. Ramsey Pricing for Two Substitute Services

stitutes. For example, suppose a cable operator responds to broadcast pay-television competition by lowering the price of pay cable. In order to maintain profits, the price of basic cable must be raised. The required price increase is higher when basic and pay cable are substitutes than when they are demand-independent. When they are substitutes, a price reduction for pay cable also reduces the demand for basic cable.

For systems that impose the technical constraint of offering pay cable only to basic-cable subscribers, the demand interdependence between basic and pay cable is complementary. That is, the cross-price elasticities are negative; reducing the price of one service increases the demand for the other, and vice versa. The technical constraint, however, burdens pay cable with a price disadvantage. A household wanting only a pay-television channel must pay the subscription price of both pay cable and the unwanted basic cable. If the sum of basic- and pay-cable were significantly higher than the price of broadcast pay television (STV), the cable operator could lose a

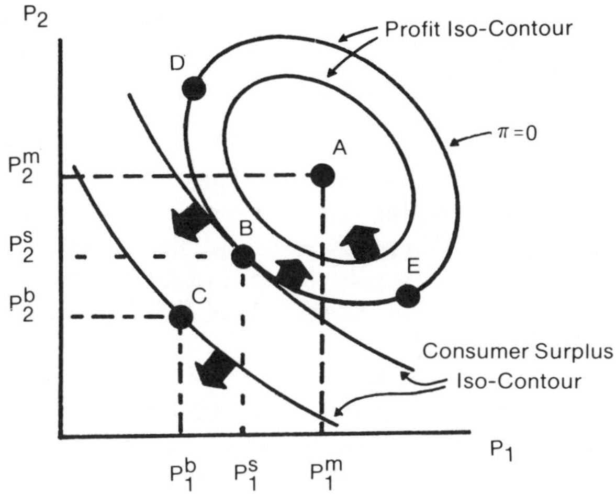

Notes:

P_1^m is the profit-maximizing price for service 1.

P_2^m is the profit-maximizing price for service 2.

P_1^b is the social-welfare-maximizing price for service 1—the "best" price.

P_2^b is the social-welfare-maximizing price for service 2—the "best" price.

P_1^r is the Ramsey, or constrained social-welfare-maximizing price for service 1.

P_2^r is the Ramsey, or constrained social-welfare-maximizing price for service 2.

A is the profit-maximizing set of prices.

B is the Ramsey, or constrained social-welfare-maximizing set of prices.

C is the social-welfare-maximizing set of prices.

D is a feasible set of Ramsey prices when service 1 is a strong complement.

E is a feasible set of Ramsey prices when service 2 is a strong complement.

Figure 5–5. Ramsey Pricing for Two Complementary Services

large number of subscribers. To avoid this loss, cable operators may elect to offer pay cable as a separate service in markets where broadcast pay television presents effective competition. In that event, nearly all services on the cable system would be substitute goods.

Some of the unusual pricing strategies employed in the franchise bidding of the early 1980s resulted from the enforced complementary relationship between basic and pay cable. The pricing strategy when services are complementary is illustrated in figure 5–5. As in the previous two figures, the characterized firm offers services having identical marginal costs and price elasticities. Given these market conditions, point *B* identifies the Ramsey prices. Both the profit and consumer's surplus contours for complementary services are less convex, more nearly straight lines, in the area of point *B* than for the case of demand-independent or substitute services. The more complementary the relationship among goods and services, as measured by a larger negative cross-price elasticity, the less convex are the profit and consumer's surplus contours. Also, the stronger the complementarity, the less inefficient is a departure from the Ramsey prices.

When goods are complements, the loss of consumer's surplus resulting from a price change is not large because the slope of the consumer's surplus and profit isocontour is nearly the same. At the extreme, when the complementarity is so strong that both profit and consumer's surplus contours are straight lines with the same slope, any set of prices that allows the firm to earn a normal profit would yield the same level of consumer's surplus and would therefore be Ramsey optimal.

Depending on the nature of the complementarity, the efficient pricing strategy may seem counterproductive. For example, points *D* and *E* in figure 5–5 identify appropriate Ramsey prices when one good strongly complements another. Point D applies when service 1, represented in the figure by price P_1, is complementary with service 2. Point *E* applies to the reverse case, when service 2 is the stronger complement. At both points one of the services is priced above the monopoly price, P_m, that would prevail if the service were produced independently! It may also occur that a price should be set below marginal cost; because these are Ramsey prices, however, no other set of prices is economically more efficient.

These results will seem startling to those students of economics who recall the emphasis placed on marginal cost pricing in standard textbooks. Historically, many economists and regulators have favored the notion that efficiency is improved as prices approach marginal cost. As the following empirical example will demonstrate, pricing efficiency is not always achieved by moving prices toward marginal cost.

Ramsey Prices in a Cable-Television Market

Like many other large urban areas during the cable-television franchising boom that began in the late 1970s, the city of Philadelphia requested proposals for cable-system construction. The market conditions that prevailed in Philadelphia during 1979, described in appendix 5B, provide the empirical backdrop for the examples of pricing and system design developed in this and the following chapter. The Philadelphia market generally reflects the conditions under which more than a hundred urban franchises were awarded during a period of four years, from 1979 to 1983. Because of inflation, differing market conditions throughout the country, and the estimation uncertainty described in earlier chapters, only general conclusions should be inferred from the following examples. Specific Ramsey prices will depend on specific market conditions.

The Ramsey prices for the following examples were calculated using a computer model, also described in appendix 5B, that combines the information on costs and revenues from chapters 3 and 4. The model forecasts pro forma income statements for a cable-television firm over the duration of the franchise contract. The expected revenues and costs for each year are adjusted for taxes and franchise fees. All figures represent 1979 constant dollars. Prices are chosen to maximize the consumer surplus from all services available on the cable system, subject to the constraint that the system earns a normal profit. As described in chapter 3, the rate of return on equity that reflects the expected normal profit for a cable-television system is, net of inflation, about 8.2 percent.

Offering no pay cable, the system described in table 5–1 earns a normal profit with a monthly subscription price for basic cable of about $7.60. This price allows the firm to break even in the sense that sufficient revenues are generated to cover all costs, including the cost of capital, the financial concept equivalent to the economic concept of normal profits. The break-even price fullfills the Ramsey-pricing rule for a natural monopoly offering one service: price equals average total cost.

The Two-Service Case: Complementary Basic and Pay Cable

Ramsey prices for a cable system that offers both basic and pay cable depend not only on the local market conditions, but also on the revenue-sharing agreement between the local cable operator and the pay-television net-

Table 5–1
Economic Characteristics of a System Offering Only Basic Cable

The market: Philadelphia Northeast Franchise, including 160,000 households.
The system: A 40-channel-capacity system offering four public-access channels, fifteen imported independent signals, the local television market, and two-way capacity.

	Basic-Cable Monthly Subscription Price	*Basic-Cable Market Penetration (%)*	*Real Rate of Return on Equity (%)*	*Consumer Surplus for One Month (thousands)*
Ramsey Price:	$7.00	30.9	6.8	$530
	$7.60[a]	29.3	8.2	502
	$8.00	28.3	9.0	483

[a]The Ramsey price to the nearest $0.05. Revenue and cost assumptions are as described in chapters 3 and 4. Documentation for the computer model, and specific assumptions, appear in appendix 5B. Prices are in 1979 dollars.

work. Modeled on the arrangement between networks and affiliated broadcast television stations, the typical revenue-sharing agreement as practiced during part of the 1970s gave the pay-television network a percentage of the gross pay-cable subscription fees. For example, the local cable operator gave the Home Box Office (HBO) pay-television network about 40 percent of the gross HBO subscription revenue earned each month by the local system. This arrangement has become less popular for a two-service system because the pricing strategy of the local cable operator is at odds with that of the pay-television network. Cable operators who were free to set prices discovered that the best strategy was to lower the pay-cable price to zero and raise the price for basic cable. Since the contract was written in terms of a percentage of gross pay-cable revenue, the local cable operator had no obligation to make any payment to the pay-television network if pay cable earned no direct revenues. Of course, with free pay cable, demand for basic cable increased. In effect, the local cable operator transferred pay-cable demand into basic-cable revenue.

For the pay-television network, the best pricing strategy under these market conditions is to set the basic-cable price to zero. Demand for pay television would increase. All revenues would come from pay television. Neither of these strategies represents Ramsey prices, however, because either the local cable operator or the pay-television network fails to earn a normal profit.

At one time, Teleprompter used the pricing strategy of evading revenue sharing by offering free pay cable for its New York cable system. This strategy, however, does not explain Cablevision's 1982 offer of free HBO as part of a franchise proposal to San Francisco. By that time HBO had rewritten the terms of the revenue-sharing contract.

Pay-television networks began writing contracts that charged local cable operators on a per-subscriber basis to discourage the evasive pricing strategies. Figure 5–6 illustrates the resulting Ramsey prices when a pay-cable channel is added to the Philadelphia cable system. These prices are based on a charge to the local cable operator of \$2.80 per month for each pay-cable subscriber, representing 1979 averages. The firm can expect to earn a normal profit when basic cable is priced at \$5.75 per month and pay cable is priced at \$6.50 per month. This price combination is represented by point *A* in the figure. The basic-cable price is almost \$2.00 a month lower than was required when no pay cable was available on the system, for two reasons. First, pay-cable revenues contribute substantially to profitability, so the basic-cable price must be lower if the firm is to earn no more than a normal profit. Second, basic cable is a stronger complement to pay cable than pay cable is to basic cable. Lowering the basic-cable price increases pay-cable revenues more than an equivalent pay-cable price reduction would increase basic-cable revenues.

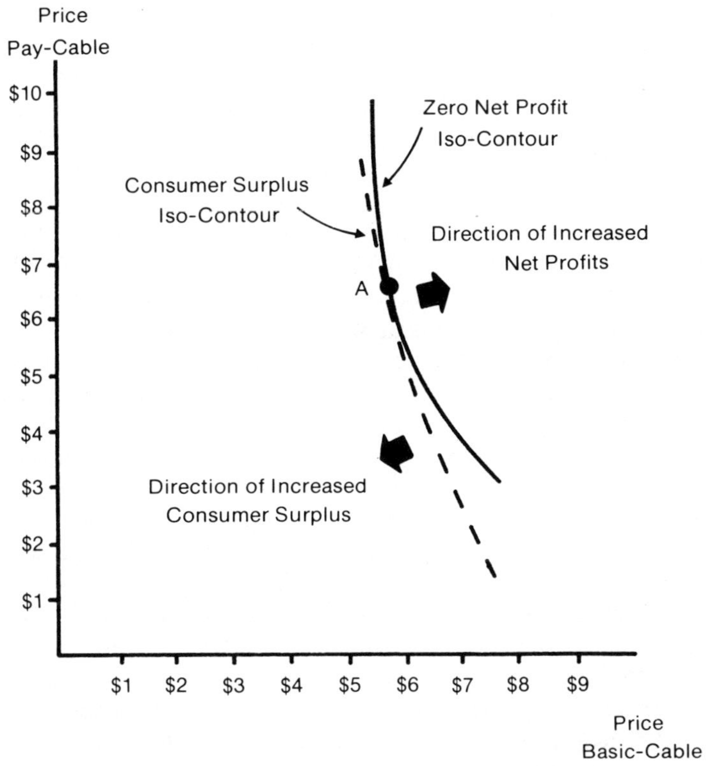

Figure 5-6. Ramsey Prices for a Basic- and Pay-Cable Service under a Charge-per-Subscriber Contract, Philadelphia Market

The suggested Ramsey prices for the Philadelphia system are somewhat lower than the national average. Economies of scale and density for a large system, as in Philadelphia, allow lower prices. Also, all systems do not offer pay cable, which reduces the required basic-cable price considerably. Nevertheless, prices could be more efficient if the charge-per-subscriber contract were abandoned.

Neglecting the terms of the charge-per-subscriber contract, the marginal cost of providing pay-cable service is small. Presumably, the marginal cost to the pay-television network is nearly zero. Pay-television networks purchase or produce most of their programming for a fixed price that does not vary on the basis of the number of viewers. Also, the equipment costs for incremental subscribers are absorbed by the local cable operator and are very small.

If the marginal costs of providing pay-cable service are nearly zero, then total profits from pay cable are achieved by choosing a price that maximizes total pay-cable revenues. In the case of a system offering only one pay-cable channel, both the pay-television network and the local cable operator should want to select the price that would maximize total profits. After all, the contract between them is just a method for dividing the total profit pie. The larger the pie, the bigger the pieces. This is the logic of the contract used in the broadcast industry, which is based on a percentage of the gross advertising revenues. Given pricing freedom, the local broadcast station will generally attempt to maximize advertising revenues. The television network, which provides most of the programming and earns a percentage of the gross advertising revenues, shares the economic goal of the local station—maximum advertising revenue. As a result, a percentage-of-the-gross contract is workable. The local station has an incentive to select the same price that the network would select.

The multiple-service nature of the local cable system defeats the economic logic of a percentage-of-the-gross contract. The cable operator may opt to evade sharing revenue with the pay-television network by setting the pay-cable price to zero and raising the price of basic cable. As an alternative, however, the charge-per-subscriber contract has the drawback of imposing an artificial marginal cost on the local cable operator. The charge-per-subscriber is artificial in the sense that it does not represent a cost, but is a means of dividing revenues—a billing formality. The local cable operator perceives this billing formality as a *real* marginal cost and raises the pay-cable price accordingly. Pricing strategies are based on the level of marginal cost, real or perceived.

The effect of the charge-per-subscriber contract is that everyone loses. Both the cable operator and the pay-television network earn less profit than could have been earned because the price of pay cable has been raised above the joint profit-maximizing level. Higher prices also mean a loss for the subscriber, measured by reduced consumer's surplus.

A different contractual agreement would encourage pricing efficiency. For example, consider a contract that specifies a lump-sum payment by the cable operator equal to the maximum revenue the pay-television network could expect to earn if revenues were shared according to a percentage-of-the-gross formula. Ramsey prices under this contract, which eliminates the artificial marginal cost, are represented by point *A* in figure 5–7. The basic-cable price under the lump-sum contract is higher than it was under the charge-per-subscriber contract, \$6.25 per month as opposed to \$5.75 per month. The advantage of the contract is that, reflecting the elimination of the artificial marginal cost, the pay-cable price is much lower, \$3.35 per month as opposed to \$6.50 per month.

The increased efficiency of the lump-sum contract from the consumer's perspective may be verified by noting the increased consumer's surplus of

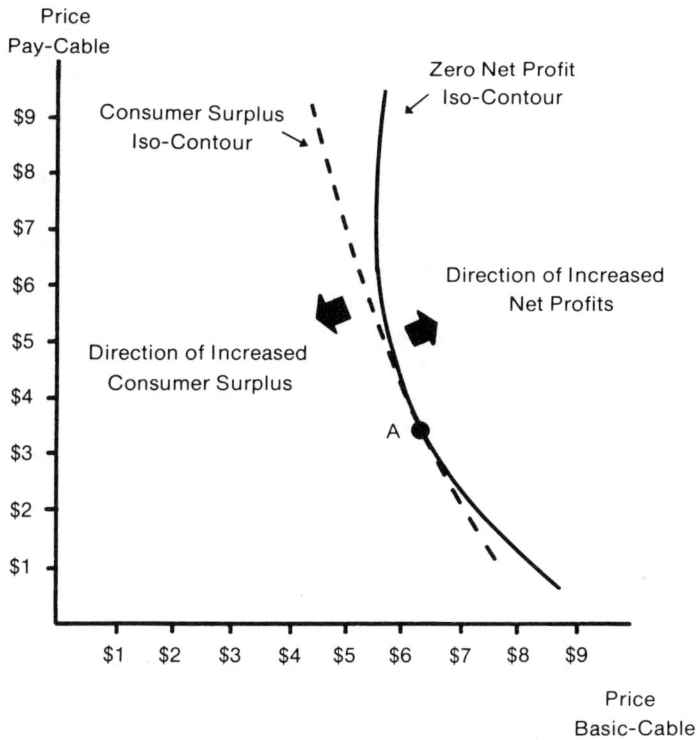

Figure 5-7. Ramsey Prices for a Basic- and Pay-Cable Service under a Lump-Sum Contract, Philadelphia Market

more than 5 percent—to $794,000 per month from $753,000 per month. That is, for the market, the lump-sum contract saves consumers about $41,000 each month. The consumer's gain with the lump-sum contract is not the result of any economic loss to the cable operator or the pay-television network. The cable operator earns the same profit as before, and the pay-television network actually receives a higher monthly revenue.

The Five-Service Case: Basic Cable and Four Pay-Cable Channels

The pricing strategy used by Cablevision to win the 1981 Boston franchise, low-priced basic cable and relatively high-priced pay cable, depends on the effects of complementarity. In order for the Boston system to break even, the low basic-cable price must result in higher-than-average pay-cable revenues. Is this strategy efficient?

The question may be answered by considering the Philadelphia market and a cable system similar to the one proposed by Cablevision for Boston. If the system offers four pay-cable channels which are priced on the basis of the charge-per-subscriber contract, then the Ramsey price for basic-cable is \$2.00 per month—exactly the same as the price bid by Cablevision for Boston. Of course, it is something of a coincidence that the prices are identical, since market conditions and system size are not the same. The Boston system also offers additional basic-cable channels at a higher price. Nevertheless, the results for the Philadelphia market, described in table 5–2, support the logic of low-priced basic cable as a bidding strategy. The results also illustrate a few generalizations for efficient cable-service pricing.

When pay cable is offered as an option to basic-cable subscribers, increasing the number of revenue-increasing pay-cable channels reduces the Ramsey price for basic cable. The price may actually fall below marginal cost. Pay-cable prices are driven toward profit-maximizing levels. If pay cable were offered as a service separate from basic cable—as a potential substitute—the pricing strategy would be based on competition between the two services rather than on the complementarity reflected in table 5–2. In

Table 5–2
Ramsey Prices for a Five-Service Cable System under Two Pay-Television Contracts

The market: Philadelphia Northeast Franchise, including 160,000 households.

The system: A 40-channel-capacity system offering four pay-cable channels, four public-access channels, fifteen imported independent signals, the local television market, and two-way capability.

Lump-Sum Contract: Payment based on a percentage of the gross; each pay-television network receives a lump-sum payment equal to 40 percent of the gross revenue that would be earned by the pay-cable channel if the subscription price were set to maximize revenue.

Service:	*Basic Cable*	*HBO*	*Showtime*	*PRISM*	*Economy Movies*
Ramsey price (\$)	2.40	7.00	7.00	9.00	4.00

Monthly consumer's surplus: \$1,378,000.

Charge-per-Subscriber Contract: The pay-television network collects a monthly charge for each subscriber to the local channel that distributes the network's programming.

Service:	*Basic Cable*	*HBO*	*Showtime*	*PRISM*	*Economy Movies*
Ramsey price (\$)	2.00	8.00	8.00	10.00	5.00

Monthly consumer's surplus: \$1,381,000.

Note: PRISM is a pay-television network operating in the Philadelphia area that programs recently released movies, local sports, and other entertainment. Economy Movies offers a limited selection of G- and GP-rated movies and is priced at a discount to the other pay-cable channels. Revenue and cost assumptions for these prices are as described in appendix 5B.

that event, the Ramsey prices for basic and pay cable would be more nearly the same.

The magnitude of pricing inefficiencies decreases as the complementarity of one of the services becomes stronger. For example, the pricing inefficiency associated with the charge-per-subscriber contract is diminished. When the system offered only one pay-cable service, consumer's surplus was more than 5 percent lower under the charge-per-subscriber contract than under the lump-sum contract. As the number of pay-cable channels increase, the complementarity of basic cable becomes stronger. Consequently, when the system offers four pay-cable channels, as in table 5–2, the difference in consumer's surplus between the two contracts was negligible.

Inefficient Pricing from Basic-Cable Rate Regulation

When systems are subject to basic-cable rate regulation, the cable operator has an economic incentive to maintain the complementarity between basic and pay cable. If the regulator limits the price of basic cable to a level that yields a normal profit for the basic part of the system, the cable operator may be able to earn excess total profits by raising pay-cable prices. As was shown in the example of pricing for Philadelphia, the efficient price of basic cable depends on the price and presence of other services. The price of basic cable should be lower when pay cable is available on the system.

The local regulator may well be expected to set basic-cable prices without regard to pay cable since local authority is legally limited to basic cable. Jurisdiction over pay-cable prices belongs to the Federal Communications Commission, but pay-cable prices are not regulated. Separating the regulatory jurisdiction appears to prevent regulators from choosing efficient prices. It also raises the practical problem of having to separate basic- and pay-cable costs when regulation is based on costs, as in the case of a local authority relying on rate-of-return regulation. Unfortunately, there is no universally accepted method of separating costs, such as system maintainance, that may be shared by both basic and pay cable.

If pay cable were offered to all households, not just to basic-cable subscribers, lower basic-cable prices would reduce potential pay-cable revenues: the services would be substitutes. As a result, the cable operator would be less able to escape the effects of basic-cable rate regulation on total profits. Also, the local regulator would be likely to set the price of basic cable too low when services are substitutes. A local regulator concerned only with basic cable ignores the loss of pay-cable profits that result from lower basic-cable prices. The cost to society from the potential pricing inefficiency may be large. Departures from Ramsey prices tend to reduce social welfare substantially when services are substitutes.

Inefficient Pricing from Competition

Introducing into the market a substitute service that is provided by another firm may also cause a departure from efficient, Ramsey prices. For example, a cable operator might reduce pay-cable prices if confronted with competition from an STV station offering pay television. With lower prices, more households may be able to subscribe to pay television, but this benefit is more than offset if the cable operator must raise the price of basic cable to maintain system profitability. Given this scenario, the result of competition is a net loss of efficiency.

The greater the scale economies of the existing firm, the greater the efficiency loss from competition. Significant scale economies appear to exist for cable television. The pricing-efficiency loss as a result of competition, therefore, may be considerable. Of course, the nonprice advantages usually attributed to competition may still apply: the cost saving of not having to support a regulatory bureaucracy and the potentially rapid technological development of unregulated firms.

Price Tiers and Pay-per-View

The economic efficiency of the cable system improved as the number of pay-cable channels increased. One reason for the improvement is that, based on the equations estimated in chapter 4, pay-television networks provide programming more valuable to the consumer than the programming of rebroadcast television stations or other independent signals. Pay-cable channels also improve the economic performance of a system because they give consumers the option of selecting from more than one level of price and service. A system offering only basic cable at a single price, a take-it-or-leave-it decision, excludes consumers who would be willing to contribute a little revenue to the system in exchange for a few channels of service. For other consumers, the market price is less than they would be willing to contribute in order to improve cable service. In general, consumers are better off when their individual demands are more nearly reflected by available price and service combinations.

As a means of improving economic efficiency by increasing the number of purchase options, some cable systems offer cable television in tiers of service. Price tiers for a hypothetical cable system appear in table 5-3. Notice that the subscription price increases less than proportionately with the number of channels available in each tier. This strategy reflects a time-honored economic rule: the law of diminishing marginal utility. It implies that the price per unit of service that consumers are willing to pay declines with the number of units consumed. Cable operators use tiers to structure

Table 5-3
An Example of Price Tiers for a Hypothetical Cable-Television System

Monthly Price	*Service*	*Service Option*
$4	Basic-cable channels 1–10	A pay-cable channel for an additional $10 per month
$7	Basic-cable channels 1–20	Two pay-cable channels, either one for $10 or both for $18 per month
$9	Basic-cable channels 1–30	Three pay-cable channels: one for $10, two for $18, or three for $24 per month
$10	Basic-cable channels 1–40	Four pay-cable channels: one for $10, two for $18, three for $24, or four for $28 per month

prices that accommodate consumer demand as closely as possible. As a result, the economic efficiency of the cable system is improved, allowing lower average prices for service, more service, and higher profits for the unregulated cable operator.

Going one step further in accommodating consumer demand as closely as possible, some systems charge subscribers on a pay-per-view basis. This pricing method allows subscribers to pay only for the programs they watch. As a result, their revenue contribution to the system closely reflects the value of service they receive, and the economic efficiency of the system is improved. This improvement will be offset to some extent by the cost of installing a more sophisticated billing system.

Some industry experts have argued that subscribers will be befuddled by the number of decisions that must be made under pay-per-view. Yet similar billing methods used in other markets—long-distance phone calls and metered household electricity or gas consumption—have historically inspired more ire than befuddlement.

Conclusions

Cable-television prices should reflect the natural-monopoly status of cable systems and the demand interdependencies of cable service. Efficient, Ramsey prices achieve the highest level of consumer's surplus while allowing a normal profit for the firm. Adding a pay-cable service to a cable system will reduce the efficient basic-cable price. For a system offering several pay-cable channels, the complementarity between basic and pay cable justifies the strategy of setting the price for basic cable at a very low level while setting the prices for pay cable at near monopoly levels.

Departures from efficient pricing impose a cost on society that tends to be more severe when services are substitutes. Examples of circumstances that may prevent pricing efficiency include: competition from noncable services, the contractual agreement between local cable operators and pay-television networks, and the separation of regulatory jurisdiction between federal and local authorities.

Note on the Appendixes to Chapter 5

Appendix 5A contains a mathematical sketch of the theory underlying the Ramsey optimal-pricing rules referred to in chapters 5 and 6. Appendix 5B contains documentation for the computer-based economic model used to develop the empirical examples of cable-system prices, profitability, and consumer's surplus that are discussed in the chapters. Much of appendix 5B is devoted to explaining the pro forma income statements that are produced by the computer model. This format was chosen for the model because municipal governments typically request pro forma income statements as part of the proposal that must be submitted by private companies participating in a round of competitive bidding to win a cable-television franchise. Also, private firms typically evaluate investment decisions in this fashion. Readers interested in more general issues of cable-television economics may skip these appendixes without loss of continuity.

Appendix 5A Optimal Pricing for a Demand-Interdependent Multiproduct Firm

The profit function for a multiproduct monopolist is given as

$$\pi(P) = X(P)P - C(X(P)) \qquad (5A.1)$$

where $\pi(P)$ is the profit function.

P is the price vector $P = (P^1, P^2, \ldots, P^n)$ for the n commodities produced by the firm.

$X(P)$ is the corresponding demand vector for the n commodities.

$C(X)$ is the cost function.

The profit-maximizing set of prices and output can be determined by differentiating the profit function with respect to price, as follows:

$$\partial\frac{\pi(P)}{\partial P} = X(P) + P\frac{\partial X}{\partial P} - \left(\frac{\partial C}{\partial X}\cdot\frac{\partial X}{\partial P}\right) \qquad (5A.2)$$

The first order conditions are obtained by setting equation 5A.2 equal to zero:

$$0 = X(P) + P\frac{\partial X}{\partial P} - \left(\frac{\partial C}{\partial X}\cdot\frac{\partial X}{\partial P}\right) \qquad (5A.3)$$

By rearranging the terms in equation 5A.3, the first order conditions for profit maximization can be written as:

$$1 = -\eta\left[\frac{P - \dfrac{\partial C}{\partial X}}{P}\right] \qquad (5A.4)$$

where

$$\frac{\partial X}{\partial P} = \begin{bmatrix} \dfrac{\partial X^1}{\partial P^1} \cdots\cdots \dfrac{\partial X^1}{\partial P^n} \\ \vdots \qquad \vdots \\ \dfrac{\partial X^n}{\partial P^1} \cdots\cdots \dfrac{\partial X^n}{\partial P^n} \end{bmatrix}$$

$$\frac{\partial C}{\partial X} = \begin{bmatrix} \dfrac{\partial C^1}{\partial X^1} \cdots\cdots \dfrac{\partial C^1}{\partial X^n} \\ \vdots \qquad \vdots \\ \dfrac{\partial C^n}{\partial X^1} \cdots\cdots \dfrac{\partial C^n}{\partial X^n} \end{bmatrix}$$

and

$$\eta = \frac{\partial X}{\partial P}\frac{P}{X}$$

The Generalized Ramsey-Pricing Rule

Traditional economic concepts suggest the social-welfare criterion of equation 5A.5:

$$W(P) = S(P) + \pi(P) \tag{5A.5}$$

where $S(P)$ is consumer's surplus, the area below the demand schedule and above the market price for each of the n services produced by the firm.

$\pi(P)$ is supplier's surplus, the area above the industry supply schedule and below the market price for each of the n services produced by the firm.

Given this formulation of the social-welfare criterion, the effects of changes in price on services or commodities not supplied by the firm are neglected.

It is typically argued that prices equal margin cost in the rest of the economy. The Langrangian first-order conditions for maximizing the social-welfare criterion of equation 5A.5 are:

$$\frac{\partial W(P)}{\partial P} - t\,\frac{\partial \pi(P)}{\partial P} = 0 \tag{5A.6}$$

where t is the Langrangian multiplier.

Ignoring the income effect of a change in price, conventional concepts of consumer's surplus yield equation 5A.7:

$$\frac{\partial W(P)}{\partial P} = -X(P) \tag{5A.7}$$

Substituting equation 5A.7 into 5A.6 results in:

$$-X(P) = t\,\frac{\partial \pi(P)}{\partial P} \tag{5A.8}$$

Inserting equation 5A.2 into 5A.8 yields:

$$-X(P) = t\left[X(P) + P\frac{\partial X}{\partial P} - \frac{\partial C}{\partial X}\cdot\frac{\partial X}{\partial P}\right] \tag{5A.9}$$

Rearranging terms gives:

$$-X(P)\frac{(1+t)}{t} = \left(P - \frac{\partial C}{\partial X}\right)\cdot\frac{\partial X}{\partial P} \tag{5A.10}$$

Letting

$$\lambda = \frac{1+t}{t}$$

$$\lambda = -\frac{\partial X}{\partial P}\frac{P}{X}\left[\frac{P - \dfrac{\partial C}{\partial X}}{P}\right] \tag{5A.11}$$

for 0 $\leq \lambda \leq 1$, which completes the development of the generalized Ramsey optimal-pricing rule. For a more thorough discussion of Ramsey-pricing theory see Rees (1968).

Appendix 5B The Economic Simulation of a Cable-Television System

The simulation model is a computer program that combines the estimates for equipment and operating costs as described in chapter 3 with the equations forecasting revenue as described in chapter 4. The program also calculates yearly tax expense, investment tax credits, interest expense, and franchise fees. A diagram of the simulation model appears in figure 5B–1.

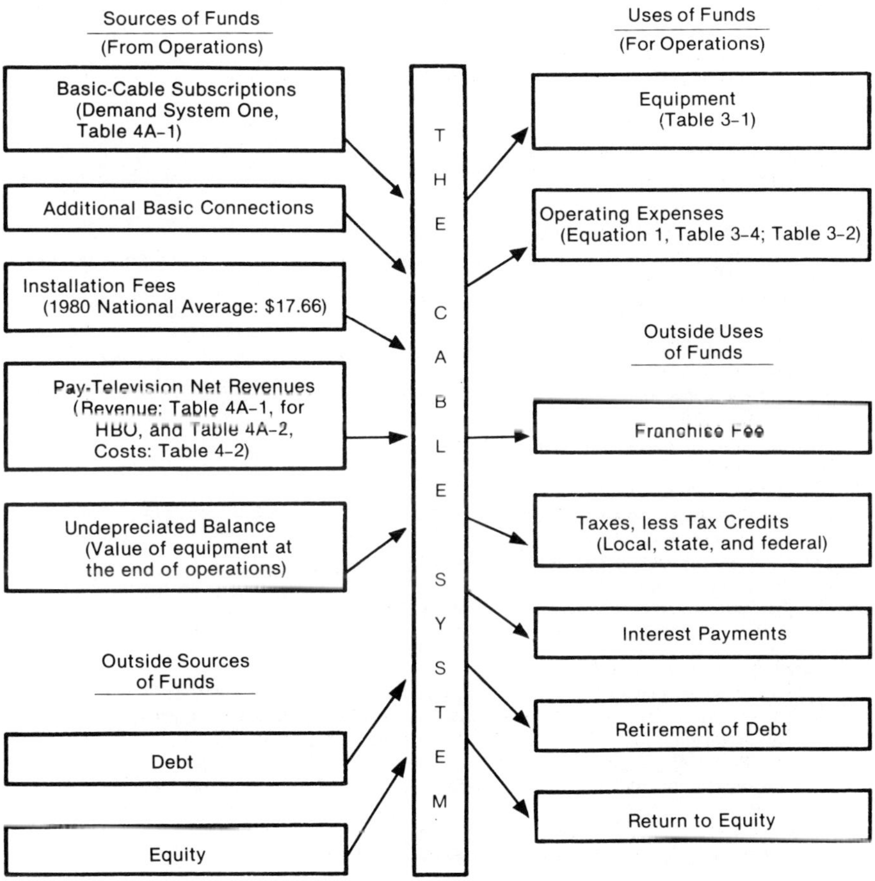

Figure 5B–1. A Cash-Flow Diagram of the Economic Model

The program is designed to accept a data file describing the characteristics of a proposed cable-television market. The data file includes demographic information and a description of the local television market, including the numbers and types of signals, and the location of the cable system relative to the B-contour of the local broadcast stations. Specific characteristics of the cable system to be evaluated—prices, numbers and types of services offered, the length of the planning horizon, the cost of capital—are input interactively. Table 5B–1 summarizes the data input into the model. A summary of the data common to the empirical examples of chapters 5, 6, and 7 appears in table 5B–2.

Table 5B–1
Data Input into the Economic Model

A. *Variables stored in an on-line data file*
1. A description of the local television market: the numbers and types of broadcast stations and their distances from the cable system.
2. A numeric description of the local terrain.
3. Socioeconomic variables, including: education, Index of Home Equipment, per capita income, and the number of households within the cable franchise area.

B. *Variables determined interactively*
1. The planning horizon: the number of years over which revenue and costs are to be forecast.
2. The cost of equity capital.
3. The interest rate on debt.
4. The monthly subscription price for cable services.
5. The installation fee.
6. The franchise fee.
7. The tax rate.
8. The number of stations imported by satellite.
9. The number of stations imported by microwave.
10. The number of locally originated channels.
11. The number of imported network stations.
12. The system channel capacity.
13. The region in which the system is located.
14. The length of plant for the system.
15. The percentage of underground construction required.

C. *Parameters stored in the program*
1. The inflation rate (1979 constant dollars unless otherwise specified).
2. The population-growth rate (zero growth rate unless otherwise specified).
3. The investment tax credit (10 percent for equipment having a useful life of seven years or more; 6.67 percent for equipment having a useful life of five or six years; 3.33 percent for equipment having a useful life of less than five years. Based on 1979 tax laws).
4. Depreciable lives for equipment (table 3–1).
5. Default values for the variables determined interactively.

Table 5B–1 continued

D. *Cost and revenue assumptions*

Cost Component	*Reference*
Capital Costs:	
Headend	Table 3–1
Distribution system	Table 3–1
Subscriber interface	Table 3–1
Other equipment	Table 3–1
Access and local programming	Table 3–2
Operating costs:	
Contractual programming	Table 3–2
Access and local programming	Table 3–2
All other	Equation 1, table 3–4
Revenue Source	*Reference*
Basic cable	Table 4A–1
Second basic-cable connection	22% of basic-cable connections
Pay TV:	
Home Box Office	Table 4A–1
Showtime	Table 4A–2
PRISM	Table 4A–2
Economy	Table 4A–2

Table 5B–2
Characteristics of the Cable Market Used in the Empirical Examples of Chapters 5, 6, and 7

Location
Philadelphia, Pennsylvania: Franchise area 4 (the Northeast).

Demographics
Households in the franchise: 160,000.
Per capita income: $4,780 (in 1974 dollars).
Percentage of adults having finished high school: 42.5%.

Local television market
Three primary network stations.
Three independent stations.
One educational station.

Technical
Miles of plant: 900 (determined by Teleprompter, Inc., from an engineering survey).
Percentage of underground plant: 2.3% (from Teleprompter, Inc.).
Note: 40-channel-capacity systems include 36 downstream (headend to subscriber) and 4 upstream (subscriber to headend) channels. One downstream channel is assumed to be used for an alphanumeric display (news or weather); 12-channel systems have no upstream capacity.

Table 5B–2 continued

Financial
Cost of equity capital: 8.2% (see chapter 3).
Cost of debt: 10.0% (1979 cost of long-term debt for Teleprompter, Inc.).
Debt as a source of funds: 48% (based on Teleprompter's debt-to-equity ratio).
Tax rate: 57.5% (including state and local taxes).
Franchise fee: 5%.

The output from the simulation model, a sample of which appears in table 5B.3, includes statistics that summarize the economic and technical character of the cable system under evaluation. The bulk of the output from the simulation model is a pro forma projection of the income statements that would be expected from the cable system over a prescribed planning horizon. Definitions for the important economic statistics and a line-by-line guide to the income statements are provided here as documentation for the simulation model.

Glossary of Terms for the Pro Forma Income Statement

DCF-ROE, the discounted-cash-flow return on equity represents the real rate of return from the equity portion of the net cash flow generated by the cable system. Equity, in this context, represents the cash flow to and from owners such as partners or stockholders. The other source of cash is debt. Unless otherwise specified, the simulation model is designed to assume that 52 percent of the cash required to fund the operation is equity, the contribution of owners. This ratio of debt equity as a source of funds represents the 1979 capital structure of Teleprompter, Inc. As developed in chapter 3, the cost of equity capital for the cable industry was estimated by using Teleprompter as a representative firm. Teleprompter's capital structure must, therefore, be used in the model because the cost of equity capital is thought to depend on the debt-to-equity ratio of the firm. Changing the debt-to-equity ratio would presumably change the cost of equity capital.

Of the positive net cash flow generated by the cable system, 48 percent is used to retire debt or make interest payments, and 52 percent is returned to equity investors. In each period a cash balance of $25,000 is maintained for contingencies. All outstanding debts are repaid in the final period, as if the cable system were liquidated.

As estimated in chapter 3, a normal profit for a cable system requires about an 8.2 percent real rate of return from equity capital. This rate of return is the cost of equity capital. The rate may be calculated from equation 5B.1:

$$I_0 = \int_0^T e^{-rt} E(t)dt \qquad (5B.1)$$

where r is the rate of return to equity.

I_0 is the initial equity investment prior to operation of the cable system.

$E(t)$ is the equity cash flow generated by the cable-system operation at time t, for the period from 0 to T.

DCF-ROI, the discounted-cash-flow return on investment, represents the equivalent continuous interest rate reflected by the net cash-flow of the cable system. The DCF-ROI reflects the entire net cash flow of the system, not just the equity portion that is included in the DCF-ROE. In equation 5B.1, r represents the DCF-ROI when $E(t)$, the equity cash flow, is replaced by $N(t)$, the net cash flow.

Net Present Value is the present value of the equity cash flow discounted continuously at the estimated cost of equity capital, 8.2 percent. The calculated net present value reported from the model is based on equation 5B.2. Ramsey prices are achieved when the net present value is approximately zero; that is, there are no excess profits.

The net present value reported in table 5B.3 of $3,495,600 for example, indicates that a firm should be willing to pay about this amount in order to win the right to build the cable system and offer services at the prices suggested in the table. Even after making this lump-sum payment of nearly $3.5 million, the cable operator would still earn a normal 8.2 percent real rate of return on equity.

$$\text{NPV} = \int_0^T e^{-.082t} E(t)dt - I_0 \qquad (5B.2)$$

where NPV is the net present value of the equity cash flow of the cable system.

Consumer's surplus is the area below the demand line and above the market price. The value reported from the model represents consumer's surplus for one month from the first period of operation. The calculation reflects a fully constructed cable system—that is, cable is available to every household in the franchise area. Also, for lack of empirical data, it was assumed that no household would be willing to pay more than $25.00 per month for any one cable service.

Table 5B-3
Pro Forma Income Statements for Philadelphia Franchise Area 4, 1979: Sample Output from the Computer-Based Economic Model
(thousands of dollars)

System Characteristics	*DCF-ROE .134*	*DCF-ROI .112*	*Net Present Value 3495.6*	*Consumer Surplus 997.*	*First Period Penetration .35238*
Period	*0*	*1*	*2*	*3*	*4*
Miles of cable		300	600	900	900
Basic penetration		.352	.391	.414	.431
Basic revenue		792.4	2,638.1	4,658.8	5,816.9
− Franchise fee		−66.5	−213.7	−365.2	−442.6
+ Installation fees		155.8	362.9	397.3	227.7
− Basic expense	−342.0	−2,347.5	−2,512.9	−3,200.9	−3,333.8
+ Net Pay-TV revenue		382.3	1,272.9	2,247.9	2,806.7
− Depreciation		−594.7	−987.5	−1,399.9	−1,607.1
Pre-tax cash flow	−342.0	−1,678.1	599.8	2,338.0	3,468.0
− Taxes (57.5%)		0.0	0.0	−504.6	−1,994.1
+ Investment tax credit		804.3	518.0	540.8	161.1
Post-tax cash flow		−873.8	1,077.8	2,374.2	1,634.9
+ Depreciation		594.7	987.5	1,399.9	1,607.1
+ Undepreciated balance		0.0	0.0	0.0	0.0
− Capital cost	−2,226.8	−5,985.4	−5,180.0	−5,408.4	−1,810.7
Net cash flow	−2,568.0	−6,264.5	−3,114.8	−1,634.3	1,431.3
Cumulative cash flow	−2,568.0	−8,833.3	−11,948.1	−13,582.4	−12,151.1

Period refers to the operating year of the cable system. The examples of chapters 5 and 6 are based on a ten-year planning horizon. Period zero, the initial investment period, represents the start-up time for the cable system. During period zero, the engineering design studies are completed, proposals are developed and filed with municipal franchise authorities, and construction is begun on the headend and office facilities.

Miles of cable is the total length of distribution plant installed by the end of each operating year. The examples are based on engineering studies conducted by firms that participated in the 1979 franchising bidding for the city of Philadelphia.

Basic penetration represents, for each year, the expected number of basic-cable subscribers as a percentage of households in the franchise area that

Table 5B–3 continued

Basic Price $7.00	*Pay-TV Price $7.00 (HBO) $7.00 (Showtime) $9.00 (PRISM) $4.00 (Economy)*	*Pay-TV Penetration .33 .12 .26 .08*	*Channel Capacity 40.*	*Local Channels 7.*	*Imported Signals 15.*
5	*6*	*7*	*8*	*9*	*10*
900	900	900	900	900	900
.444	.455	.464	.472	.479	.485
5,994.0	6,139.5	6,263.0	6,370.2	6,464.9	6,549.7
−446.1	−456.5	−465.5	−473.2	−480.1	−486.3
34.8	28.6	24.3	21.1	18.6	16.7
−3,218.2	−3,095.3	−2,970.9	−2,839.1	−2,851.3	−2,862.2
2,892.2	2,962.4	3,021.9	3,073.7	3,119.4	3,160.3
−1,627.0	−1,658.4	−1,675.7	−1,687.7	−1,698.4	−1,711.3
3,629.8	3,920.3	4,197.1	4,464.8	4,573.0	4,666.9
−2,087.1	−2,254.2	−2,413.3	−2,567.3	−2,629.5	−2,683.4
23.1	26.8	26.4	14.0	12.4	21.7
1,565.8	1,692.9	1,810.2	1,911.5	1,955.9	2,005.1
1,627.0	1,658.4	1,675.7	1,687.7	1,698.4	1,711.3
0.0	0.0	0.0	0.0	0.0	7,666.5
−231.0	−306.7	−471.0	−139.8	−123.6	−430.7
2,961.8	3,044.6	3,014.8	3,159.4	3,530.7	10,952.3
−9,189.3	−6,144.7	−3,129.9	329.5	3,860.2	14,812.5

have access to cable. The forecast is based on the equation for basic cable in table 4A–1.

Basic revenue is the annual cash inflow from subscriber fees for the basic-cable system, including fees for additional basic connections, but not including installation fees.

Franchise fee represents a local tax levied by the municipality. Although the legal maximum, determined by the FCC, is 3 percent of gross revenues, waivers are often granted to allow for a 5-percent franchise fee.

An **Installation fee** of $16.58, the 1979 national average, is collected once from each subscriber to the system.

Basic Expense includes basic and local programming costs, salaries, and other system-operating expenses. Figures for basic expense are based on equation 1 in table 3–4 and table 3–2. Also included is interest on debt held by the local cable system. A 10-percent interest rate is assumed in the model, reflecting Teleprompter's 1979 rate on long-term debt. Also in period zero, a start-up expense of about $242,000 is expected, the average for systems proposed in Philadelphia during 1979. This amount is within FCC-suggested guidelines. Start-up expenses include noncapitalized labor such as legal fees and other costs associated with the preparation of cable-franchise proposals.

Net pay-TV revenues are annual pay-cable revenues net of the contractually determined payment that the local cable operator makes to the pay-television network. Revenues are also net of a handling fee of $1.25 per subscriber per year to cover billing expenses and program guides. Gross pay-television revenues are based on the equation in table 4A–1 for HBO, and on the equations in table 4A–2 for other pay-cable services.

Depreciation is calculated using the straight-line method of equation 5B.3:

$$\text{Depreciation} = \sum_{i=1}^{c} \frac{\text{Cost}_i}{N_i} \tag{5B.3}$$

where, Depreciation is the yearly depreciation expense.

Cost_i is the original cost of the ith piece of capital equipment being depreciated in each period (table 3–1).

N_i is the useful life of each piece of equipment (table 3–1).

c is the number of pieces of capital equipment being depreciated in each year.

It is assumed that all equipment is replaced at the end of its useful life and there is no salvage value.

Pretax cash flow represents the sum of basic revenue, installation fees, and net pay-TV revenues; less franchise fees, basic expense, and depreciation. The pretax cash flow represents the cash flow that the local cable system would report for tax purposes given the straight-line-depreciation assumption. Accelerated depreciation will increase profitability.

Taxes—including state, local, and federal taxes—amount to 57.5 percent for the examples based on the city of Philadelphia, but will vary somewhat among cities. A tax-loss carry-forward is used in the model; that is, period losses may be carried by the company for up to seven years to be applied against future positive profits.

Investment tax credit (ITC) is an after-tax credit granted to the taxed entity by the federal government. Tax credits are assumed to be used by the parent company in the year they are awarded, even if the cable system generates no tax liability during the period. The ITC is calculated as a percentage of the total cost of capital equipment. The percentages prevailing at the beginning of the simulation period, 1980, were: 10 percent for capital with a useful life of seven years or more; 6.67 percent for capital having a useful life of five years or more but less than seven years; and 3.33 percent for capital having a useful life of less than five years.

Posttax cash flow is the pretax cash flow plus the investment tax credit and less taxes.

Depreciation does not represent a real cash flow but is subtracted from revenues for tax purposes. To offset this deduction for tax purposes, depreciation is added back into the posttax cash flow.

Undepreciated balance represents the book value of equipment, original cost less depreciation, at the end of the planning horizon. The undepreciated balance is added into the posttax cash flow to represent the value of equipment remaining at the end of the planning horizon. Of course, book value merely represents an estimate of the true value of equipment in the cable system.

Capital cost represents the cash outlays required to purchase and install cable-system equipment. All equipment is replaced at the end of its useful life. These costs, detailed in table 3-1, are the actual cash-flow represented for tax purposes by depreciation.

Net cash flow is the posttax cash flow plus depreciation and undepreciated balance, less capital cost. In the final period, all outstanding debts are repaid.

Cumulative cash flow is the sum of the current and previous periods' net cash flows.

6 Cost-Benefit Analysis of System Design

Each cable system is designed to accommodate the demands of a specific market. The design process involves decisions about channel capacity, programming capability, and system length, as well as more technical decisions regarding the choice of electronic components. Described in the language of cost-benefit analysis, the best design provides the highest level of benefits net of costs. From the perspective of a decision maker in the private sector, benefits may be strictly defined as revenue, in which case the best design results in the highest profit.

Given the conditions of pure competition, market forces will encourage firms to allocate resources efficiently and so to choose the best design. Nevertheless, design decisions require an interpretation of market forces. Cost-benefit analysis is a method of presenting information about the market to a decision maker; it includes a projection of costs and benefits for each design under consideration. The pro forma income statement, described in appendix 5B, is a popular format for cost-benefit analysis in the private sector. Pro forma income statements are also typically required as part of the proposal a private firm must submit to government authorities when bidding for a cable-television franchise. In the franchising process, private firms are invited to submit proposals in competition. The winning design is decided not by market forces, but by the government. Once a design is accepted, direct competition is all but eliminated from the market.

The Role of Consumer's Surplus in Cost-Benefit Analysis

If the government authority has economic goals, it should select the cable system design that maximizes the social-welfare criterion—that is, the sum of consumer's and supplier's surplus (equation 5.1). Given the economies of scale that characterize cable-television systems, supplier's surplus should be at a level that guarantees a normal profit for the cable operator.

If the private firms participating in the franchising process perceive that the government has economic goals, they should propose system designs that maximize the social-welfare criterion. Wanting to have their proposals selected, private firms will choose a design that conforms to the perceived goal of the agency awarding the franchise. Private firms have the additional incentive of choosing the design that maximizes the welfare criterion because it would be the most profitable system if all consumer's surplus could be converted into revenue. Consumer's surplus is potential revenue. As discussed in chapter 5, pay-per-view pricing strategies are one method of converting consumer's surplus into revenue.

In his classic text on cost-benefit analysis, E.J. Mishan (1973) notes that "consumer's surplus is the most crucial concept in the measurement of social benefits in any social cost-benefit calculation." To illustrate the concept, consider a government authority that has received three proposals to build a cable-television system for a municipality. A graphical cost-benefit analysis of the proposed systems appears in figure 6–1.

The systems are ordered in the figure on the basis of channel capacity and programming capability. System A is the most restricted design, offering fewer channels and less programming than systems B or C. As a result, the total cost of building and operating System A is the lowest for any of the three proposed designs. The subscription price for System A, P_a, which represents the cost of delivering service to each subscriber, is not the lowest price among the three systems, however. Because System A offers only a few programming services, not many households are willing to become subscribers; demand is low. With relatively few households to share the fixed costs of the system—a significant portion of cable-television total costs—the price must be kept high if the cable operator is to earn a normal profit.

System B in figure 6–1 reflects a proposed cable system design that includes more programming services than the System A design. The better comparative service of System B results in a higher level of demand. With more subscribers to share in the cost of the system, service can be offered at a lower price even though the total cost of building and operating System B is greater than the total cost of System A. Households receive more service at a lower price. In the language of cost-benefit analysis, the higher cost of building System B, compared with System A, is more than offset by the difference in benefits. Net social benefits, as measured by consumer's surplus, are higher for System B. It is the better design.

Of the three proposed designs, System C offers the most channels of programming, or programming of a higher quality. Consequently, the demand for System C is greater than the demand for Systems A or B. Even allowing for the higher demand, however, the cost of building and operating System C requires the highest subscription price if the cable operator is to earn a normal profit.

The choice of the best system design is less intuitive when System C is included as an alternative. Among the three systems, C offers the best service, but B is the lowest in price. The choice of system design depends on how society values the additional service of System C.

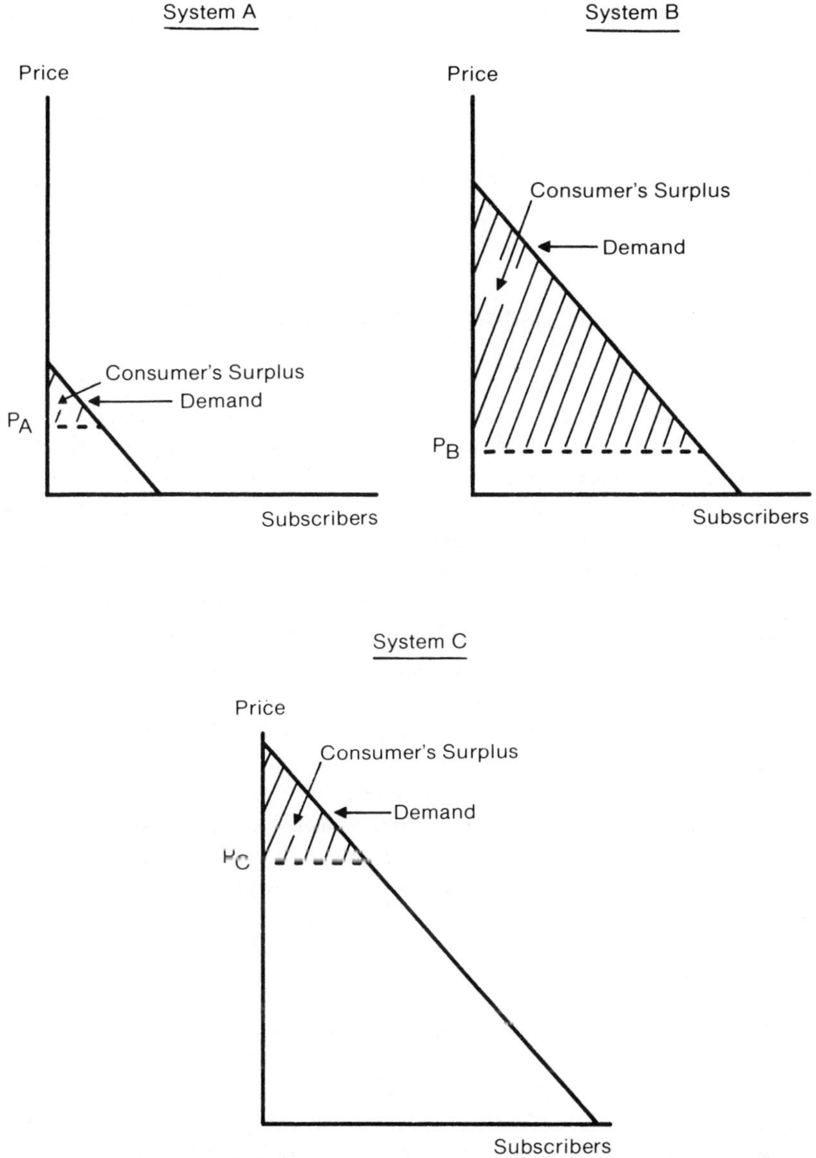

Figure 6–1. The Cost-Benefit Analysis of Three Cable Designs

The graphical cost-benefit analysis in Figure 6–1 indicates that the benefits of more programming associated with C are more than offset by the higher cost. Net social benefits, the difference between benefits and cost measured by consumer's surplus, are highest for System B. It is the best design of the three.

In this example the biggest is not the best. System C is too expensive, at least for the static market. Changing conditions of cost or demand may alter the results of the cost-benefit analysis. Anticipating these changes—the dynamic aspect of the market—is one of the most challenging problems encountered in cost-benefit analysis.

The Telecommunications Tradition of Universal Service

According to economic theory, the use of consumer's surplus or similar measures of benefits is appropriate for the public evaluation of designs for a project such as a cable-television system. Other measures of benefits have also been used, however. Until recently, the telecommunications tradition of providing universal service had been widely applied to public decisions regarding telephone service. Deregulation has shifted responsibility for some of these decisions to the private sector. For local telephone service, which is still subject to extensive public regulation, the goal has been to provide universal service to the extent that it is economically feasible. In other words, the goal is to maximize the number of users subject to the normal profit constraint. The U.S. Postal Service operates with similar goals.

The goals of social welfare and universal service will often support the same design selection when alternatives are limited. Among the three systems represented in figure 6–1, System B best fulfills the goals of both social welfare and universal service. It would produce the most consumer's surplus and attract the largest number of subscribers. In general, however, social welfare favors somewhat more expensive designs than does universal service, giving more weight to quality of service.

A Case Study Using Cost-Benefit Analysis: Cable Television for Philadelphia

A first step in the cost-benefit analysis is determining the efficient price for each system design to be evaluated. Efficient prices convey information about the cost of each design and are used to determine the value of services offered by the system. The examples of efficient pricing for cable services in chapter 5 were based on market conditions that prevailed in Philadelphia during the 1979 round of franchise bidding. The Philadelphia market typifies the urban markets in which most of the franchise action has been focused in recent years.

The first set of cable designs to be evaluated in this section are for a system with 900 miles of distribution plant serving 160,000 households. In the second part of this case study, the problem of choosing the best market size will be considered. System length will be allowed to vary in order to investigate the comparative benefits of enlarging the cable-system market area.

The particulars of the local market are described in appendix 5B as part of the documentation for the computer-based financial model used to calculate efficient prices and consumer's surplus. Systems are evaluated with respect to the level of comparative programming service that would be available to subscribers. Improving the comparative service may require the use of more cable channels, but coaxial cable is available in only a limited number of sizes. Until the early 1980s, cable was most commonly available in 12- and 40-channel capacities. An 80-channel system, therefore, required two 40-channel coaxial cables.

The cost-benefit analysis begins with the most rudimentary cable design and generally proceeds in order of increasing comparative programming service. In a sense, the order is also historic. The earliest cable systems were, of course, the most rudimentary. As more programming services became available, larger systems were built. The market forces at work in the development of the cable industry are evidenced in the economic evaluations conducted in the following cost-benefit analysis.

The evaluation of each cable-system design includes an estimate of efficient prices for cable services. Efficient prices maximize consumer's surplus subject to the constraint that equity investors earn at least an 8.2-percent real rate of return. The level of consumer's surplus reported with the evaluation of each system reflects one month of operation and includes both basic-cable and, when available, pay-cable services. According to the social-welfare criterion, the design that produces the most consumer's surplus is best.

System Design One, which has a 12-channel capacity and distributes the local television market, produces about $179,000 worth of consumer's surplus each month. Recall that consumer's surplus measures the value of service in excess of the monthly subscription price, of $10.25 per subscriber in the case of System Design One. Table 6–1 summarizes the system design characteristics.

The low expected market penetration, 11.5 percent, is attributable to the limited service advantage that System Design One offers compared with the local broadcast market. Since Design One would distribute nothing more than signals available from off the air, only households that receive low-quality broadcast signals would be potential subscribers. Given the proximity of local stations and the relatively flat local terrain, most households probably receive an acceptable broadcast signal. Households located

Table 6–1
System Design One

Programming: The local television market
Channel capacity: 12
Basic-cable subscription price: $10.25 per month
Basic-cable market penetration: 11.5% (first year)
One month of consumer's surplus: $179,000

Note: The efficient, break-even price is estimated to the nearest $0.25 using the computer model documented in appendix 5B. Prices are in 1979 dollars. The system would be 900 miles in length, serving 160,000 households in Northeast Philadelphia.

near tall buildings or power lines, and therefore experiencing poor reception, might consider paying the high monthly subscription price. Because of the inelastic demand for basic cable, the cable operator could not lower the subscription price and still earn a normal profit. If System Design One were offered at $7.00 a month, about the national average price, return on equity would decline from about 8.2 percent to about 2.7 percent. The system would still be in the black, but equity investors could do better by putting their money into money-market funds.

Location is the problem for System Design One. If constructed in a market more distant from the local broadcast stations, the system would be more profitable. After all, the first cable system that was built by John Watson is essentially System Design One, but located sixty miles to the north. Market penetration was higher for the first system because it had the advantage of offering good-quality broadcast signals to households whose location prevented adequate signal reception from off the air.

Imported Signals

The FCC's decision that allowed cable operators to import distant broadcast signals for distribution in local markets had a dramatic impact on the development of cable. No longer were cable operators constrained to providing the service of improved reception. The decision was especially important for cable systems in urban areas, where reception of off-the-air signals was usually quite good. This economic significance is evidenced in the improved market penetration, lower break-even price, and higher level of consumer's surplus for System Design Two—a 12-channel system distributing the local television market in addition to five distant broadcast signals imported by microwave (see table 6–2). In 1966, cable systems similar to

Table 6–2
System Design Two

Programming: Five microwave-imported distant independent broadcast signals, and the local television market

Channel capacity: 12

Basic-cable subscription price: $7.50 per month

Basic-cable market penetration: 19.5% (first year)

One month of consumer's surplus: $318,000

Note: The efficient, break-even price is estimated to the nearest $0.25 using the computer model documented in appendix 5B.

Design Two were proposed for the city of Philadelphia. As in many cities, FCC regulatory restrictions relating to distant signals from the mid-1960s to the mid-1970s discouraged development of these early urban systems.

Part of the advantage of Design Two, compared with Design One, is that nearly all of its 12 channels of capacity are filled with desirable programming. Design One uses only about half its 12 channels. Nevertheless, subscribers to Design One must contribute enough cash in the form of fees to finance the purchase of the 12-channel-capacity coaxial cable, not readily available in a smaller size. Capital costs for both systems are about the same, so the marginal cost of adding five imported signals is quite low. The economic advantage of not having to finance excess capacity is sometimes referred to as *economies of fill.*

The use of land-based microwave to import distant signals is limited by the expense of boosting signals every twenty or thirty miles to overcome blockage from terrain or the curvature of the earth. With the advent of satellite distribution, the number of broadcast signals that could be economically imported increased. Satellites also fostered an explosive growth in the number of independent broadcast networks, such as Cable News Network or Satellite Music Network. These events encouraged larger cable systems, as evidenced in the economic characteristics of System Design Three, which includes all the programming services of Design Two in addition to ten satellite-imported signals (see table 6–3). Design Three draws a higher market penetration and yields about 24 percent more consumer's surplus than Design Two, despite its slightly higher price. Design Three is better; it has more potential profit.

In part, the slightly higher price of Design Three reflects the economic inefficiency of an underutilized 40-channel system. In the early 1980s many of the 40-channel systems around the country were operating with excess channel capacity.

Table 6–3
System Design Three

Programming: Ten satellite-imported distant independent signals, five microwave-imported distant independent broadcast signals, and the local television market

Channel capacity: 40 (with two-way capability)

Basic-cable subscription price: $8.00 per month

Basic-cable market penetration: 26.7% (first period)

One month of consumer's surplus: $395,000

Note: The efficient, break-even price is estimated to the nearest $0.25 using the computer model documented in appendix 5B.

Local Origination

As a source of programming for the cable system, local origination includes both public-access programming and locally produced programming. Once required by FCC regulations, a local-origination requirement is now often included by municipal authorities as a condition for granting franchise awards to private companies. Following the form of the previous FCC regulations, many municipal authorities require that systems allow for public access to four channels, one each for: (1) the government, (2) the schools, (3) the community, and (4) leased access. The economic logic of these requirements has been widely debated.

System Design Four accommodates four public-access channels in addition to all the programming services of Design Three. The evaluation of Design Four, summarized in table 6–4, indicates that, at least for urban systems, public-access programming can improve the economic characteristics of the system. The increased demand that results from the additional four public-access channels allows the cable operator to offer basic-cable service at a lower price; it also yields about 27 percent more consumer's surplus. Given these market conditions, even an unregulated cable operator would have an economic incentive to offer public access.

The cable operators who are more likely to suffer an economic hardship as a result of providing simple public-access equipment are those with small systems located in rural areas. With fewer subscribers to share the mostly fixed cost of public-access equipment, these operators must raise prices significantly if they are to earn a normal profit. Since the average market-penetration rates for rural systems are much higher than for urban systems, the marginal revenue that can be expected from a price increase—and the increased demand associated with providing public access—will be lower for rural systems. For many, the economic benefits will not offset the

Table 6-4
System Design Four

Programming: Four public-access channels, ten satellite-imported distant independent signals, five microwave-imported distant independent broadcast signals, and the local television market

Channel capacity: 40 (with two-way capability)

Basic-cable subscription price: $7.60 per month

Basic-cable market penetration: 29.3% (first period)

One month of consumer's surplus: $502,000

Note: The efficient, break-even price is estimated to the nearest $0.05 using the computer model documented in appendix 5B. In the estimation it was assumed that any revenues directly generated by these channels, such as leasing fees, are used by the cable operator to provide better equipment than the historical average assumed in the computer model.

costs. Acknowledging the differential impact of public access, the FCC imposed fewer requirements on very small rural systems when it enforced public-access requirements.

The economic controversy surrounding public access on large urban systems has more to do with the level of investment encouraged by the government authorities who award cable-television franchises. Private firms participating in the competitive bidding to win the franchise awards have, in many cases, made a commitment to provide elaborate public-access equipment. The required investment may run into the millions of dollars. Taking the notion of public access one step further, municipal authorities have required that a separate cable, with 40 or 50 channels of capacity, be dedicated to the needs of the local government—an institutional network, as it is called.

The institutional network may serve as a video telephone or data-transmission system connecting schools, police, or other municipal operations. Typically the institutional network is provided to the municipality free of cost.

In some cases the cost of the institutional network has been absorbed by the cable operator, who, overestimating potential revenues in the market, earns less than a normal profit. More often, cable operators pass the cost along to subscribers in the form of higher fees; in effect taxing the subscribers to finance equipment used by the municipality. Understandably, government authorities have found that this indirect method of taxation—higher cable prices—meets with less public resistance than raising property taxes. The higher cable fees have not been perceived as a tax by the public.

Another local use of the cable system has been the programming of

Table 6–5
System Design Five

Programming: One locally programmed channel, four public-access channels, ten satellite-imported distant independent signals, five microwave-imported distant independent broadcast signals, and the local television market

Channel capacity: 40 (with two-way capability)

Basic-cable subscription price: $8.25 per month

Basic-cable market penetration: 29.1% (first period)

One month of consumer's surplus: $498,000

Note: The efficient, break-even price is estimated to the nearest $0.05. In the estimation, no advertising revenues were forecast for the locally programmed channel.

channels by the cable operator. System Design Five, described in table 6–5, provides for a locally programmed channel as well as for all the programming services offered by Design Four. The economic evaluation assumes that no advertising revenue is earned by the locally programmed channel. Though long anticipated as a signifcant revenue source, advertising sales accounted for less than 1 percent of total revenues during the early 1980s. Without advertising revenue, the addition of a locally programmed channel modestly erodes the system's economic performance. Both consumer's surplus and market penetration are somewhat lower for System Five than for System Four, which offers public access but no locally programmed channel.

The economic disadvantage of local programming could become an economic advantage, depending on the cable operator's ability to earn advertising revenue from the locally programmed channel. For the cost-benefit analysis, this revenue should be counted net of any loss in subscriber fees from households who prefer not to view commercial advertisements. The net advertising revenue would enable the cable operator to offer basic cable at a lower price and thus increase consumer's surplus. Firms choosing to advertise on the system would also account for some additional consumer's surplus—the difference between the market price of advertising and the price firms would have been willing to pay.

For locally programmed channels on an individual system, the benefits do not appear to offset the costs without advertising revenue. Accordingly, many cable operators have been reluctant to produce local programming. Of course, local programming is good public relations. Some local programming is motivated by the desire to provide the community with an electronic public forum and by the allure of participating in show business. Often a commitment to local programming is a requirement of the franchise agreement written by municipal authorities, who in practice do not always pursue economic goals. Locally programmed channels are offered on 90 percent of the cable systems.

Pay Cable

Much of the reluctance of cable operators to develop local programming is due to the significantly better economic opportunity afforded by pay television. For example, replacing the locally programmed channel of System Design Five with a pay-cable channel results in more than a 50-percent increase in consumer's surplus. The system with the pay-cable channel, Design Six, is the better design. It offers the cable operator the potential for larger profits. If the price of basic cable were raised to $7 per month, somewhat less than the national average, Design Six would yield a return on equity of more than 16 percent—twice the normal 8.2 percent profit.

Part of the consumer's surplus and potential profits are lost if the pay-television network and local cable operator cannot agree on an efficient contract to share revenue. Design Six is the system used in the previous chapter to illustrate Ramsey pricing for two interdependent services. The Ramsey prices that appear in table 6–6 as part of the economic evaluation of Design Six are based on the assumption that the economically efficient lump-sum contract defines the relationship between the local cable operator and the pay-television network.

Adding three more pay-cable channels to the system, making System Design Seven, produces another dramatic improvement in economic performance (see table 6–7). Design Seven, with four pay-cable channels, yields over 40 percent more consumer's surplus than Design Six, with one pay-cable channel. Again, Design Seven assumes that the more efficient lump-sum contract is in force.

The Static Optimum System Design

Of the systems so far evaluated in the cost benefit analysis, Design Seven is the best as measured by consumer's surplus. In keeping with the telecommunications tradition of providing universal service to the extent possible, Design Seven would bring service to more households than any other system design. It allows for sufficient channel capacity to distribute the major satellite and pay-television networks that were available during the 1979 franchise bidding for Philadelphia.

Design Seven would be the best for the franchise area under two conditions: a mature market with no expectation of growth in services or change

Table 6–6
System Design Six

Programming: One pay-cable channel distributing Home Box Office, four public-access channels, ten satellite-imported independent signals, five microwave-imported independent signals, and the local television market.

Channel capacity: 40 (with two-way capability)

Basic-cable subscription price: $6.25 per month

Basic-cable market penetration: 33.7% (first period)

Pay-cable subscription price: $3.75 per month

Pay-cable system penetration: 54.8% (of basic cable)

One month of consumer's surplus: $794,000

Note: Ramsey prices for the two services are calculated with the assumption that a lump-sum contract is negotiated between the local cable operator and pay-television network. Prices are estimated to the nearest $0.05.

Table 6–7
System Design Seven

Programming: Four pay-cable channels, four public-access channels, ten satellite-imported independent signals, five microwave-imported signals, and the local television market

Channel capactiy: 40 (with two-way capability)

Basic-cable subscription price: $2.40 per month

Basic-cable market penetration: 51.9% (first year)

Pay-Cable Service	*HBO*	*Showtime*	*Prism*	*Economy*
Pay-cable price ($):	7.00	7.00	9.00	4.00
Pay-cable penetration (%):	33	12	26	8

One month of consumer's surplus: $1,378,000

Note: Ramsey prices are calculated assuming a lump-sum contract. The basic-cable price is estimated to the nearest $0.05, pay-cable prices to the nearest $0.25.

in demand, and stable technology, with no expectation of cost-reducing innovations. Design Seven was the *static optimum*—the best design if time would bring no appreciable changes to the economic evaluation. Of course, as is now obvious, time would bring many changes to the industry.

The Dynamic Optimum

During the years that the city of Philadelphia evaluated the proposals it received in 1979, cable systems with as many as 100 channels were proposed for other smaller cities. The largest system proposed for Philadelphia contained only 66 channels devoted to programming. Many local citizens felt that Philadelphia should also have a 100-channel system. In part it was a question of prestige.

The danger of building too large a system is that there will not be enough suitable programming to fill all the channels. As previous examples have demonstrated, maintaining excess channel capacity erodes economic performance. The result is higher-than-necessary prices.

Except for underground construction and the subscriber interface, there is little cost advantage in anticipatory construction. That is, the cost of installing two cable lines is about the same whether the lines are installed at the same time or sequentially. Another consideration is that the technologically dynamic nature of the cable industry creates a potential cost disadvantage for anticipatory construction. By overbuilding with state-of-the-art technology, cable operators forego the opportunity of expanding capacity when improved technology becomes available. This opportunity cost is especially high for a rapidly developing technology such as cable. With some thoughtful planning, system capacity can be expanded as required. The system should be designed, however, so that it is not underbuilt when it becomes operational. For a system of the size of Design Seven, planning and construction will require three years or more.

From 1979 to 1983 the number of satellite-distributed signals increased from about 15 to about 60. The economic advantage to be gained by anticipating this market growth is evidenced in the economic evaluation of System Design Eight (see table 6–8). With 80 channels of capacity and 60 imported signals, Design Eight generates about 20 percent more consumer's surplus than Design Seven. Although the monthly subscription price of Design Eight is about $1 per month higher as a result of the expanded channel capacity and service, it is the better design.

To compensate the cable operator for the risk of adding channel capacity in anticipation of the market, a profit margin slightly higher than the normal 8.2 percent is allowed. The monthly subscription price allows for an expected 8.8-percent real rate of return on equity. Adding more channel capacity to the overall design would only erode the economic performance of the system. Design Eight can accommodate most of the major programming services scheduled for availability through the mid-1980s. For the market, Design Eight is the *dynamic optimum*—the best design allowing for expected growth.

In general, systems should be able to accommodate short-term growth, but with minimal excess capacity. The precise amount of excess capacity that should be built into the system depends on the expected growth in the demand for new services. Because of the larger potential market, urban systems will experience faster growth rates than rural systems and therefore should be designed with more excess capacity. Also, because of the large potential cost savings to be realized from the anticipatory construction of

Table 6–8
System Design Eight

Programming: Fifty-five satellite imported signals, four pay-cable channels, four public-access channels, five microwave-imported signals, and the local television market

Channel capacity: 80 (with two-way capability)

Basic-cable subscription price: $3.50 per month

Basic-cable market penetration: 57% (first year)

Pay-Cable Service:	*HBO*	*Showtime*	*Prism*	*Economy*
Pay-cable prices ($):	7.00	7.00	9.00	4.00
Pay-cable penetration (%):	33	12	26	8

One month of consumer's surplus: $1,652,000

Note: Prices are estimated to be to the nearest $0.25. Because the proposed level of programming significantly exceeds the range included in the sample used to construct the model, a high degree of uncertainty accompanies the price estimates. As compensation to the cable operator for bearing this uncertainty, a higher profit is allowed. The cable operator earns an 8.8-percent real rate of return on equity, rather than the normal 8.2 percent.

underground cable, cable operators should allow for more excess capacity when cable is underground than when cable is aerial.

Both rapidly developing technology and a changing regulatory environment increase the risk of anticipatory construction. Cable operators who overbuild with coaxial cable will be at a cost disadvantage as fiber-optic technology is developed. A legislative decision restricting cable from competing in the data-transmission market would be an economic setback for cable firms that have already added channel capacity in anticipation of the burgeoning market.

One reason for designing systems to accommodate only short-term growth is the difficulty of anticipating events in the more distant future. Nevertheless, municipalities usually grant cable-franchise awards that remain in effect for fifteen years. Because renegotiating the franchise is problematic, municipal authorities must therefore emphasize long-term planning, complicating the design problem. A few systems, designed during the early 1980s under the influence of explosive growth in the number of satellite services, will have excess capacity through the decade as the growth of satellite services moderates. Subscribers to these systems will have to pay higher prices for service than if less ambitious designs had been approved.

System Length and Market Area

One aspect of system design that has been neglected so far in the cost-benefit analysis is the market area to be served by the system. In practice, the market area is often determined by municipal boundaries because the franchising of systems is largely a municipal function. Understandably, individual municipal governments prefer to make the decisions about the type of cable system which will be built within their jurisdictional boundaries. Occasionally municipal, county, and state governments share these decisions with the idea of defining a market area based on economic, rather than political, objectives.

Perhaps the most important economic relationships to be considered in defining a cable-television market area are that the cost of service per subscriber tends to decline as both system length and population density increase. These relationships are partly attributable to the headend. As system length and population density increase, there are more subscribers to share the fixed cost of the headend. Spreading the cost among more subscribers means that each will be obliged to pay less.

The frequent amplification required to maintain the strength of signals traveling through coaxial cable limits the extent to which the cost of the headend can be shared. Excessive amplification reduces signal quality to below acceptable levels. As a result, the largest market area that can be adequately served by a single headend is about six miles in radius. In order to distribute the cost of the headend among the largest number of subscribers, therefore, each cable system should be about six miles in radius, with a headend located at the center. The geographical packing of systems that best fulfills the six-mile-radius criterion is illustrated in figure 6–2; hexagonally shaped market areas minimize the required number of headends. Improvements in amplifier technology will increase the maximum radius of each hexagonal market area. Ideally—if there were no variation in local

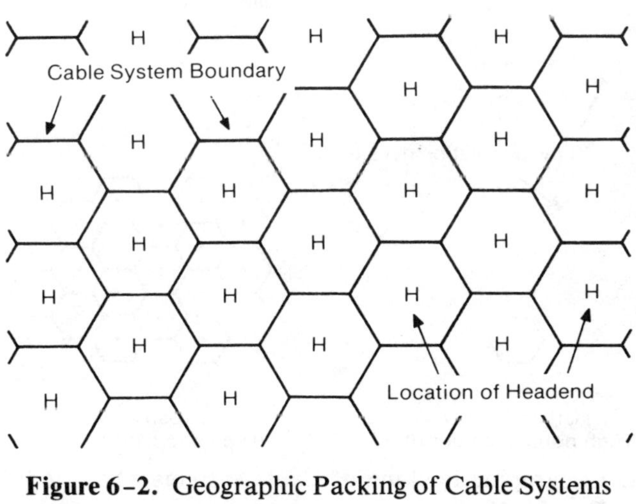

Figure 6–2. Geographic Packing of Cable Systems

conditions such as terrain, real-estate values, or population density—the country would be honeycombed with hexagonally shaped cable systems.

Conditions of local terrain may require that market areas adapt to the landscape, distorting the honeycomb pattern. For example, in rural areas it may be cheaper to build around a mountain than to go over the top; in urban areas the high cost of real estate may be a crucial factor in choosing the headend location.

Many urban areas are separated by rural areas with population densities too low to support a cable system. Figure 6–3 is an illustration of the clumped honeycomb pattern that would be expected if the only local variation in conditions were population density. In general, population density declines with distance from the center of a city.

The cost disadvantage of lower population densities at a distance from the city is offset to some extent by the operating economies for larger systems. As the length of the system distribution plant increases to serve a larger market area, the operating cost per mile decreases. The combined

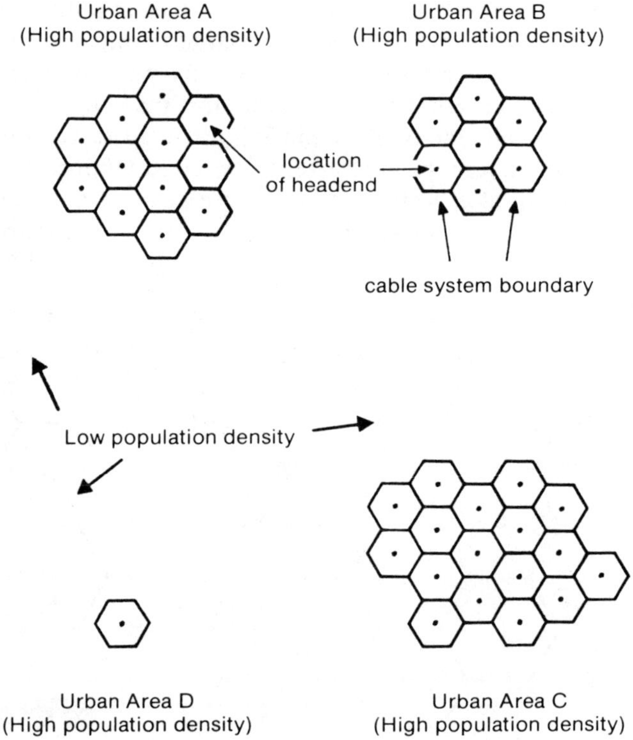

Figure 6–3. Geographic Packing of Cable Systems when Population Density Varies.

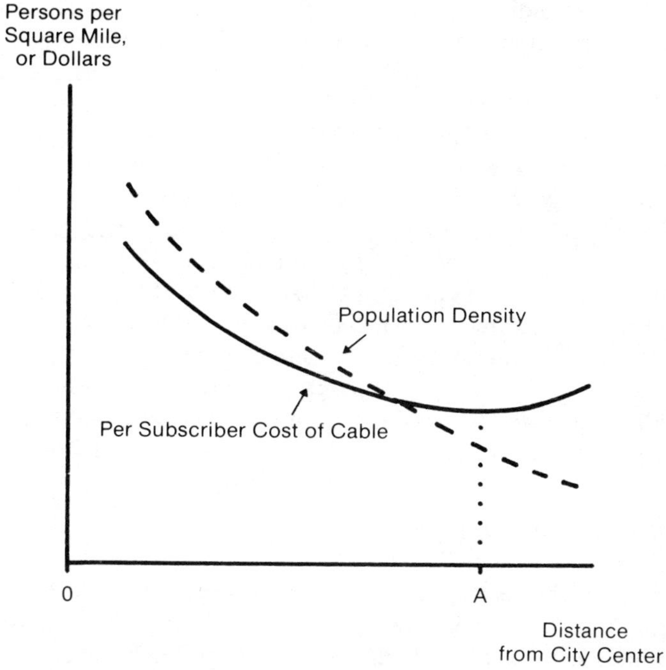

Note: 0 to A is the radius of the least-cost cable system.

Figure 6-4. The Effect of Population Density and System Length on Average Total Cost

effect of economics of density and length is illustrated in figure 6-4. The dashed line in the figure reflects the expected decrease in population density associated with increasing distance from the city center, point 0. The solid line, which reflects the cost of service per subscriber, also begins to decline as distance from the city center increases because the system is getting larger. The horizontal axis measures the radius of the cable system. At point *A,* however, the lower population density offsets the economies associated with system size; further expansion of the market area served by the system would increase the average cost of service. The distance from 0 to *A* in figure 6-4 is, therefore, the radius of the least-cost system. Point *A* is the system boundary.

The residents of households located just beyond the boundary of a cable system may be angered if they do not have the opportunity to receive service but their neighbors do. As a result, drawing the boundary may involve a politically sensitive decision. An economic solution to this dilemma is to charge households beyond the least-cost radius a higher subscription fee to reflect the higher cost of delivering service.

The least-cost-radius for a cable system need not be six miles—the radius that can be served by a single headend. Headends can be linked by microwave or super-trunk to form multiple headend cable systems. The largest system included in the sample used to build the econometric model described in chapter 3 was about 2,500 miles in length. Up to that length, operating costs per mile of distribution plant declined as system length increased. It is likely that operating costs per mile are lower for even larger systems. From a cost perspective, therefore, it would be wise to have all the systems in an urban area operated by a single firm. For example, a single firm would operate all the systems in Urban Area A of figure 6–3. Likewise, the systems in each of three other urban areas would be organized into three large operations. This organization resembles that of the telephone industry, where local service in a large geographic area is provided by a single firm. The disadvantage of this organization is that ownership becomes concentrated within a few firms, and small firms are excluded from the market.

The Citywide Optimum Design

As an example of the economic benefits to be achieved from increasing the scale of operations, consider again the case of Philadelphia. Municipal authorities had decided to divide the city into four market areas, each to be served by a separately franchised cable system. Without examining the benefits to be achieved by enlarging the market area, it was determined that System Design Eight best suited the market as anticipated through the early 1980s. A similar system would be constructed and operated by a different firm in each of the other three market areas of the city.

The cost disadvantage of using four nonoverlapping systems to serve the city is apparent in the comparison presented in table 6–9. Combining four market areas into one lowers the price of basic cable from about $3.50 per month to about $1.00 per month. Consumer's surplus for the city increases by nearly $1 million per month. Based on the cost-benefit analysis, the best design appears to be the citywide system, Design Nine. The system offers the same services as Design Eight; but by serving a larger market area, Design Nine more effectively exploits the natural-monopoly character of cable systems. Table 6–10 presents a summary of all the designs evaluated in the cost-benefit analysis. For purposes of comparison, consumer's surplus for Design Nine is calculated to represent only the part of the city that would be served by the other system designs.

Summary

The growing availability of microwave- and satellite-distributed programming services has substantially increased the value of cable television service

Table 6-9
A Comparison of System Designs Eight and Nine

System Design Eight: A separate cable system in each of four nonoverlapping market areas in the city; average system length, 588 miles

System Design Nine: A citywide system operated by a single firm; total system length, 2,327 miles

Programming (both designs): Fifty-five satellite-imported signals; four pay-cable channels (HBO, $7.00 per month; Showtime, $7.00 per month; PRISM, $9.00 per month; Economy, $4.00 per month); four public-access channels; five microwave-imported signals; and the local television market

Channel capacity: 80 (with two-way capability)

System Design	*Average Basic-Cable Price ($)*	*Average Basic-Cable Penetration (%)*	*Total Monthly Consumer's Surplus for the City ($)*
Eight	3.50	59	6,033,000
Nine	1.00	84	7,006,000

Note: Prices are estimated to the nearest $0.25. Three headends are required for the citywide system, Design Nine. There are 560,000 households in the total market; the four smaller markets were divided as follows: area 1, 130,000; area 2, 130,000; area 3, 140,000; and area 4, 160,000.

Table 6-10
Results of the Cost-Benefit Analysis, Philadelphia Northeast Franchise
(1979 Dollars)

Design	*Programming*	*Channel Capacity*	*Basic Cable Price (Dollars per Month)*	*Monthly Consumer's Surplus (Thousands of Dollars)*
One	The local television market.	12	10.25	179
Two	Five microwave imported broadcast signals, and the local television market.	12	7.50	318
Three	Ten satellite-imported independent signals; five microwave-imported broadcast signals; and the local television market.	40	8.00	395
Four	Four public-access channels; ten satellite-imported distant signals; five microwave-imported broadcast signals, and the local television market.	40	7.60	502
Five	One locally programmed channel; four public-access channels; ten satellite-imported independent signals; five microwave-imported broadcast signals; and the local television market.	40	8.25	498

Table 6–10 continued

Design	*Programming*	*Channel Capacity*	*Basic-Cable Price (Dollars per Month)*	*Monthly Consumer's Surplus (Thousands of Dollars)*
Six	One pay cable distributing HBO ($3.35 per month), four public-access channels; ten satellite-imported independent signals; five microwave-imported signals; and the local television market.	40	6.25	794
Seven	Four pay-cable channels (HBO, $7.00; Showtime, $7.00; PRISM, $9.00; Economy, $4.00), four public-access channels; ten satellite-imported independent signals; five microwave-imported signals; and the local television market.	40	2.40	1,378
Eight	Fifty-five satellite-imported signals; four pay-cable channels (HBO, $7.00; Showtime, $7.00; PRISM, $9.00; Economy, $4.00); four public-access channels; five microwave-imported signals; and the local television market.	80	3.50	1,652
Nine	A citywide system; otherwise the same programming as Design Eight	80	1.00	2,001[a]

[a]Consumer's surplus is calculated to represent the 160,000 households in the Northeast Philadelphia franchise.

and continues to provide an incentive for the construction of systems with large channel capacities. The economic value of locally produced programming is more uncertain; generally local programming requires the support of advertising revenue if it is to be cost-justified. A few low-cost public-access facilities, however, seem to represent an economic contribution to urban systems. Excessive investment in public access will negate this contribution.

The limitation on the distance signals can travel through coaxial cable would, by itself, encourage a honeycombed pattern of cable-system boundaries. The economies of scale associated with operating the cable distribution plant are responsible for the cost advantage of allowing a single firm—a monopoly—to serve a large market area.

7 The Municipal Franchising of Cable Television

The franchise is a contract that authorizes an individual or firm to market specific goods or services in a particular territory. It is a form of market organization used in businesses as diverse as fast foods and cosmetics. For example, the franchising of restaurants in the McDonald's fast-food chain is intended to give each restaurant manager a market area large enough to sustain profits. Among other benefits, franchising prevents excessive competition among individual McDonald's restaurants.

Private firms such as McDonald's rely on the legal protection of patents and trademarks to enforce franchise contracts and to prevent unauthorized market activity. Municipal governments rely on their jurisdiction over public rights of way to enforce cable-television franchise contracts. Except in a few states where municipal jurisdiction has been preempted by state government, a franchise contract must be negotiated with the municipal authorities before cable can be installed. Other utilities such as telephone, water, and electrical services must also secure consent to use public rights of way.

More than 10,000 separate cable-television franchise contracts have been negotiated between cable operators and local-government authorities during the last thirty years. These negotiations have been the subject of intense public debate, sometimes resulting in years of litigation. In many cases the franchising of cable television represents the single most important and controversial decision to be made by an administration of the local government. During the course of this stormy franchising activity, a commonly accepted procedure for the municipal franchising of cable television has evolved.

The Franchising Procedure

Four sets of documents are ordinarily prepared in the course of awarding a cable franchise: an enabling ordinance, a request for proposals, the franchise proposals, and the franchise contract.

The *enabling ordinance,* prepared by the municipality, formalizes the authority of the municipality to govern cable franchises and describes the franchising rules, such as application procedures and eligibility requirements for firms wishing to construct and operate a cable system. Many of the provisions that will be included in the franchise contract are specified in the enabling ordinance. These provisions may include the applicability of price regulation, the duration and exclusivity of the franchise contract, the technical and programming requirements for the cable system, the franchise fees to be paid, and the market area to be served.

The application procedure specified by the enabling ordinance usually involves a public round of bidding by qualified private firms. Qualifications for firms can include proof of having the financial and technical resources necessary to build and operate a system; occasionally the qualifications include minority or local participation in ownership of the system. The purpose of the ownership qualification is to ensure a community voice in management. Few local firms, however, can raise enough capital to construct a cable system. A common practice is for major corporations to offer part ownership to local citizens in order to meet the ownership qualification. Although the local citizens are usually given 20 percent of the ownership, they often become silent partners. This arrangement, popularly known as rent-a-citizen, defeats the intended purpose of the ownership qualification.

The timing and form of the competitive bidding for a franchise are determined by the *request for proposals* (RFP) drafted and distributed by the municipality. The RFP describes the requirements set forth in the enabling ordinance and specifies the application deadline and materials that should be included in the franchise proposals.

Firms wishing to compete in the bidding will respond to an RFP with separate proposals. A large part of each proposal is devoted to a technical description of the system. Each proposal also contains a descriptive list of the programming services to be distributed on the system, including the level of commitment to public access and locally originated programming. This information is usually accompanied by a schedule of prices for cable services, justified by a set of pro forma income statements. These statements, described in appendix 5B, project system revenues, costs, and profitability.

After a period of evaluation, negotiations begin between the prospective cable operators and the municipal authorities. If all goes as planned, the details of the contract are settled and the franchise is awarded. Professional consultants are often hired to assist municipal authorities in these details.

During the franchising process, public hearings are held to encourage citizen participation in the decision making. The public hearings are an effective method of ascertaining community needs. Also, one of the most important lessons of the franchising experience is the need to maintain an open franchising process.

Franchising Activity

The terms of some of the major franchise contracts awarded in the early 1980s are summarized in Table 7–1. Cablevision Incorporated, winning an exclusive 15-year franchise for Boston, caused a stir in the industry by bidding a basic-cable price as low as $2.00 per month. It was the first time a major cable firm had adopted this strategy, although smaller firms have made similar proposals in the past. For example, in 1970 Focus Cable proposed a basic-cable price of $1.75 for the city of Oakland. The country's largest cable company, American Television and Communications (ATC), recently went one step further in Pittsburgh by proposing to offer a limited number of basic-cable channels free of cost. Although ATC did not win the Pittsburgh franchise, the strategy has since been used successfully by Cox in Omaha and Rogers in Portland. Nevertheless, low-priced basic cable has not become the standard strategy. In the 1982 bidding for Sacramento, proposed prices for basic cable ranged from $1.50 to $19.50 per month. Sur-

Table 7–1
Major Franchise Awards, 1980–1982

Company/City	*Cost of Basic Cable*	*Channel Capacity*	*Households in Franchise Area (Thousands)*	*System Length (Miles)*	*Capital Cost (Millions of of Dollars)*
ATC					
Indianapolis	$6.50	42	140	1,052	24.3
Cablevision					
Boston	$2.00–8.00	108	250	750	95
Cox					
Tuscon	$6.92–13.95	108	135	1,060	38.9
New Orleans	$7.95–11.95	108	220		
Omaha	$0.00–10.95	108 (80 in use)	125	1,095	33.7
Sammons					
Ft. Worth	$3.95–10.95	54	160	1,430	50.0
United					
Scottsdale	$5.95–11.95	54	50	400	19.0
Warner Amex					
Cincinnati	$3.95–10.95	138 (86 in use)	161	1,525	25.0
Dallas	$2.95–9.95	108	400	2,360	41.6
Pittsburgh	$5.35–9.45	80 (60 in use)	181	700	

Source: Adopted from Peter Falco, senior industry specialist, Merrill Lynch, *Cable Investment.* (New York, 1983):14. Reprinted with permission.

prisingly, the high bid was submitted by Cablevision only a year after its record low bid in Boston.

While the industry has been experimenting with low-priced basic cable, the channel capacity for proposed systems has been steadily increasing. At the beginning of the decade, 40 channels of capacity was not an uncommon proposal. Within a few years systems were proposed with capacity in excess of 100 channels for home subscribers and an additional 40 or 50 channels for municipal use in the form of an institutional network.

Companies have gone to great lengths to fill the channels promised in their franchise proposals. To win the New Orleans franchise, Cox promised 18 public-access channels and a $3.38-million-dollar grant to support ethnic, cultural, educational, and general local programming. Additionally, the company promised to construct and maintain a 50-channel institutional network to be used by the city. Even so, the Cox proposal is in no way unusual; it merely reflects the remarkable concessions private firms have had to make in order to win cable-franchise contracts.

Often the concessions include cash payments by the private firm to the municipality. For example, the Tribune Company won the franchise for Tampa, Florida, with the promise of establishing a trust fund for community projects. The company's contribution would amount to an initial $1.5 million and an additional $250,000 for each year of the franchise life.

Not all franchises have been awarded to private firms. After a series of conflicts with Warner-Amex, the city of Valparaiso, Florida, built a municipally funded system. Municipally owned or operated systems have also been considered for Cleveland, Ohio; Palo Alto, California; St. Paul, Minnesota; and Paterson, New Jersey. The trend, however, has been away from municipal ownership, although many cities have experienced difficulty in enforcing the terms of the franchise contract once it has been awarded to a private firm.

In many cities like Philadelphia or Atlanta, where cable franchises were awarded in the 1960s or 1970s, the cable system was never fully constructed, and large parts of the cities were not served. More recently Cox has been reluctant to begin construction on the impressive system it had proposed in New Orleans, and the city has threatened to sue. Unfortunately, the threat of legal action has not always been sufficient to encourage fulfillment of the franchise contract. Although the municipal franchising of cable television is likely to continue as the dominant regulatory process, it is not without problems.

The Debate over the Franchising Process

The practice of awarding franchise contracts through a competitive bidding process is intended to result in the efficient pricing and design of a cable system. Economists propose that when large numbers of suppliers bid noncollusively to offer a service characterized by increasing returns to scale, the resulting price will not reflect monopoly power (Dementz 1968). Critics of the franchising process, however, assert that franchise bidding does not ensure efficient pricing and that the uncertainties associated with a developing technology like cable television make it impossible to achieve a fair franchise agreement (Williamson 1976). Much of the controversy in the literature has focused on the duration of the franchise agreement.

In the early days of cable television, franchise contracts were awarded for life—once for all. These contracts have the obvious disadvantage that beneficial changes in the industry accrue to the franchise holder on a monopoly basis unless such changes are anticipated. Once-for-all contracts that specify future prices resulting from uncertain future events are, it is generally agreed, impossible to write.

One approach to this problem, favored by Harold Dementz (1968), relies on incomplete long-term contracts that allow for renegotiation of the terms of the franchise when unanticipated events occur. Oliver Williamson (1976) describes the following problems associated with incomplete long-term contracts:

1. The initial award criterion is apt to be artificial or obscure.
2. Execution problems in price-cost, in performance, and in political respects are apt to develop.
3. Bidding parity between the incumbent and prospective rivals at the contract renewal is unlikely to be realized.

One interesting attempt to define a workable renegotiation procedure, developed by Ingo Vogelsang and Jorge Finsinger (1979), relies on the construction of a price index that limits a regulated private or public multiproduct enterprise to prices that would have resulted in normal profits if applied to the previous year. Vogelsang and Finsinger show that the short-term profit-maximizing firm will choose prices that are optimal in a social-welfare context; that is, the prices maximize the sum of consumer's and producer's surplus.

The drawbacks to this procedure, acknowledged by the authors, include that:

1. Demand elasticities and costs must remain constant over time.
2. The index is applicable only to one-period profit-maximizing firms.

In addition, the authors do not address the problem of using such an index during inflationary periods. Given inflation and a pricing criterion based on a previous year's economic performance, it is unlikely that the firms would ever achieve normal profits—always failing to meet inflating costs with prices derived from a base period. An extensive literature discusses the problem of index measurements (Afriat 1977).

The requirement that demand elasticities remain constant over time makes the Vogelsang-Finsinger approach difficult to apply to a new technology like cable television, where service innovations can be expected to affect demand. It is also clear that investors in cable-television systems are not one-period profit maximizers, since new cable systems typically generate several periods of losses before cash flows become positive. One-period profit maximizers would be attracted to other investments.

An alternative to incomplete long-term franchise agreements is recurrent bidding—short-term franchise contracts. With this scheme the initial franchise award would be granted for a relatively short time, four or five years. Prior to the end of the initial franchise agreement, firms would be invited to make competitive bids for the opportunity of operating the cable system through the next franchise period, another four or five years; subsequent franchise contracts would be awarded competitively.

Richard Posner (1972), argues that the recurrent bidding allows for sequential decision making, thereby avoiding the hazards that beset long-term contracts. Recurrent bidding, however, depends on the assumption that parity among bidders at the contract-renewal interval is realized—that is, the firm operating the cable system at the time of the contract renewal has no significant advantage in the bidding process over outside firms seeking to obtain the franchise award. Williamson (1976) argues that because of the specialized nature of cable technology, the incumbent firm is certain to have an advantage in the contest since it has better information about the value of the equipment in place. Posner (1969), on the other hand argues that this problem can "be solved by including in the franchise a provision requiring the franchisee, at the successor's option, to sell the plant (including improvements) to the successor at its original cost, as depreciated." One drawback to the recurrent bidding strategy is the high cost of preparing proposals, which can be as much as 10 percent of one year's operating and capital expense. Frequent bidding can be expected to raise significantly the cost of operating a system. The high cost of frequent bidding is one reason that most contracts are for fifteen years.

The use of long-term, nonexclusive franchise contracts is a compromise approach that has been adopted by many municipalities. This approach avoids the high cost of frequent, formal bidding but reserves the municipal government's privilege of calling for a round of competitive bidding if the performance of the incumbent cable operator is not adequate. The threat of entry from outside firms strengthens the position of the municipality when it becomes necessary to renegotiate the terms of a contract in order to accommodate changes in technology or community needs. Also, the threat of competition should discourage cable operators from raising prices very much above competitive levels. In practice, however, the specific nature of the cable operation and the high cost of competitive entry have discouraged outside firms from submitting competitive proposals against an incumbent cable operator.

Another general criticism of the franchising process concerns the selection of a system design. A common complaint is the emphasis that has been placed on public-access and institutional-network facilities. In particular, prices for service must be raised if cable operators are to finance institutional networks. Consequently, cable will be less price competitive with other technologies that are not burdened with the responsibility of extensive investment in equipment dedicated to government use. Although the municipality benefits from state-of-the-art technology, it is unclear that investments in institutional networks represent an efficient use of resources. If tax dollars instead of cable revenues were used to finance these investments, few institutional networks would be built. The franchising process, it is argued, encourages overbuilding by providing financially strapped municipal administrators with ready access to a large source of specialized investment capital.

One more aspect of the franchising process that may favor overbuilding is the attention that has been focused on system channel capacity. In the quest to achieve a state-of-the-art system, the appropriateness of a design for the individual market can be forgotten. A municipal decision maker who errs on the side of choosing an overbuilt design can portray the system as a venture into the technological frontier. Perhaps only the cable operator and a few individuals closely involved with the decision will realize that the overbuilt system will result in unnecessarily high prices. The public, of course, is likely to blame the cable operator for high prices.

Another factor that tends to encourage the choice of an overbuilt system is the element of prestige associated with an extravagant system. Sensitive to their public image, elected officials may abandon purely economic goals in their choice of design.

The Scope of the Franchising Contract

The natural-monopoly character of cable systems is the economic justification for the franchising process. A single firm can deliver service at a lower price than can any combination of competing firms. How many different services should be delivered by the single firm, however? For example, could basic cable be delivered by one firm and pay cable by another?

Two firms could be used, but there is an operating-cost advantage in allowing a single firm to deliver both basic and pay cable in a single market area. Operating costs rise less than proportionately with the number of channels of programming distributed by the system. Accordingly, it is less expensive for a single firm to distribute the services than for a combination of firms to do so. For that reason, it is sensible to franchise a market area so that a single firm offers both basic and pay cable because prices can be set at lower levels.

Defined by equation 7.1, the cost advantages that may be gained by producing more than one good within a single firm are called *economies of scope*. Economies of scope exist only if there are costs common to two or more products within the firm.

Economies of scope:

$$C(X_1 \ldots X_m) < C(X_1, 0, \ldots, 0) + C(0, X_2, 0 \ldots, 0) + C(0, \ldots X_m) \tag{7.1}$$

where C denotes cost and X denotes products 1 through m.

For example, it is cheaper to offer basic- and pay-cable services on the same system because common construction, equipment, and maintenance costs need not be duplicated. It is not clear, however, that there are costs common to both distributing services and producing programming. A lack of common costs would indicate that there is no economic advantage in having the cable operator produce the local programming. A separate firm or firms could perform that function equally well.

It has been argued that in order to minimize the concentration of ownership in the industry, the programming function should be separated from the distribution function. This separation would, perhaps, allow more firms to participate in an industry increasingly dominated by corporate giants. In practice, local programming is sometimes included as a responsibility of the cable operator by the terms of the franchise contract because otherwise no local programming would be provided. Until advertising as a source of revenues becomes more fully developed, local programming on cable will continue to be unprofitable.

A broad issue in the structure of the telecommunications industry is the economies of scope that may exist between cable and telephone services. If it is substantially less expensive to have all wire services—telephone, cable, and data transmission—provided by the same firm, should the local cable companies be combined into the newly organized local telephone companies? To date, legislation and regulations at all levels have discouraged the combination of cable, telephone, and other wire services into a single monopoly firm.

A Case Study of Franchising for Philadelphia

The first cable system, built for Mahony City in 1948, distributed signals that originated from Philadelphia. Not until 1964, however, was a cable system proposed for Philadelphia itself. By 1983 only a small section of the city had actually been wired for cable. Over the years, the Philadelphia city government has experienced most of the major problems involved in the franchising process.

The first proposal to build a cable system for the city was submitted by Triangle Publications, owned by media tycoon Walter Annenberg. At the time, November 1964, the Annenberg family owned two of the city's newspapers (the *Philadelphia Inquirer* and the *Daily News*), a Philadelphia television station (WFIL-TV), and a radio station. Although concern was voiced over this concentration of communications ownership, the major opposition came from the principals of a UHF television station about to go on the air in Philadelphia.

The Philadelphia City Council decided to decline the first Annenberg proposal, which envisioned a single large cable system servicing the entire city. One reason was that the proposal had not been formally requested and so had not passed through a competitive bidding process. In 1966, under the direction of Isadore Bellis, chairman of the law and government committee, the city council initiated a formal request for proposals. The round of competitive bidding which followed may have only served as cover for an arrangement based on the original Annenberg proposal. According to Milton Shapp, former Pennsylvania governor and one time owner of Jerrold Electronics, there was "a deal all wired to give the entire Philadelphia franchise to Walter Annenberg" (*Philadelphia Magazine* 1979). Learning of the arrangement on the evening before the final hearings, Shapp and some of his friends organized a number of franchise proposals of their own. The next morning, the city council was confronted with a large contingent of lawyers and members of the local business community. The council finally awarded six nonoverlapping cable televison franchises to the companies listed in table 7–2. Isadore Bellis, chairman of the cable committee, was later convicted of selling his influence to a food service organization that sought a franchise in Philadelphia.

The lack of public involvement in the franchising of many cities during this period encouraged the pursuit of special interests at the expense of community goals. In an interview published in the 13 August 1973 issue of the *Philadelphia Daily News,* former mayor Richardson Dilworth commented that the enabling ordinance that governed the 1966 franchising of the city had been passed "without public hearings or any consideration for the general welfare." He also noted that the only effective limitation contained in the ordinance, that franchise ownership could not be transferred, had been violated by at least one of the participating companies.

Table 7-2
1966 Philadelphia Franchise Awards

Jerrold Electronics, owned by Milton Shapp
Telesystems, the only system to be constructed; eventually taken over by Communications Properties, a Texas conglomerate
Comcast, a small local company
Philadelphia Cable TV, eventually taken over by Telesystems
The *Bulletin,* a family-owned Philadelphia daily newspaper
Triangle Publications, a conglomerate owned by Walter Annenberg and family

On 8 October 1971 Jerrold Electronics sold all of its cable systems, both operating and nonoperating, to National Transvideo, a subsidiary of Sammons Communications. Although no construction had yet begun on the Jerrold Electronics cable franchise in Philadelphia, the franchise was included in the sale with a reported value of $75,000.

Of the six companies that won the franchise award in 1966, only Telesystems actually began construction. By 1974, Telesystems was extending its cable system outside of its franchise boundaries in south Philadelphia. Aerial cable was being strung in the west Philadelphia area, franchised in 1966 to the Philadelphia Cable Television company. Since the terms of the franchise contract did not set a date by which the system must be operational, the west Philadelphia franchise still legally belonged to Philadelphia Cable Television, not Telesystems. The construction was halted when a senior-citizens group in west Philadelphia protested the stringing of overhead wires.

In 1975 the city hired the consulting firm of Malarkey and Taylor, an affiliate of Warner Communications, to make recommendations about cable franchising. The consulting firm concluded that the city should proceed with new plans to wire every neighborhood. By 1979 the city was ready for another round of competitive bidding. The six franchises of 1966 were consolidated into four franchises for 1979. A map of the 1979 franshise areas appears in figure 7-1.

The 1979 franchising of Philadelphia is typical of the municipal franchising process as it is currently practiced. The most important terms of the proposed franchise contract appear in table 7-3.

During the week of 15 May 1979, a series of meetings were held to answer questions of prospective bidders for the group franchise areas. Applications were due on 18 July, and the review and examination of bids were to last from 17 July to 15 October 1979. The bids are summarized in table 7-4.

Speculation mounted in the city's communication media as to when and in what form cable television would be introduced to Philadelphia, but no apparent action was taken by city government until 17 May 1981. On that

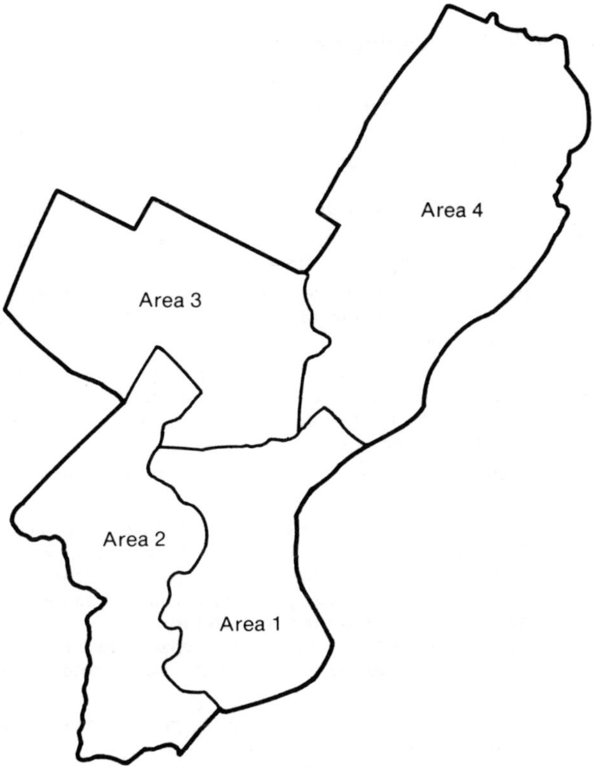

Figure 7–1. Cable-Television Franchise Boundaries for Philadelphia, 1979

day, city councilman James J. Tayoun introduced an ordinance that would transfer the power of awarding cable-television franchises away from the mayor's office to the council.

The vote on the ordinance to transfer power was to come on 26 June 1981. The day before the scheduled vote, Mayor William Green announced that the proposals of all ten companies competing for the four franchises would be rejected and that the city would, for the third time, initiate the bidding process to award the franchises. The city council agreed to delay voting on the ordinance to transfer franchising authority. The reasons given by the mayor for rejecting the bids included the rapidly changing cable-television technology, the difficulty of providing a uniform system to the city, and the political infighting that had characterized the franchising process. The mayor also announced that in the next round of bidding, each private firm winning a franchise award would be required to build and maintain an institutional network dedicated to municipal use.

Table 7–3
Terms of the 1979 Philadelphia Franchise Contract

Duration:	Fifteen years (nonexclusive).
Channel capacity:	A minimum of 30 channels must be made available. In addition, the system "shall be capable of instituting two-way operations by the addition of plug-in return components." Further, "additional channel capacity shall be made available when excessive utilization requires it."
Access channels:	A public, governmental, educational, and leased-access channel must be provided by each system.
Interconnection:	Each system must be designed to allow interconnection with the systems of other franchises, to allow simultaneous carriage of access programming.
Technical performance	Technical performance of the systems are governed by the rules and regulations of the Federal Communications Commission. Modifications of these rules must be adopted by the franchise within one year of their proposal by the FCC.
Transfer of franchise:	Franchises granted by the city are not transferable without the prior approval of the city council.
Charges to subscribers:	A uniform rate schedule will be established by the Department of Public Property for all cable subscribers in the city. "Such rate schedule may be amended by the Department from time to time upon request of any franchise provided that notice and an opportunity to be heard is provided to the public and to the nonrequesting franchises. No increase shall be approved by the Department until it has conferred with and received the recommendation of the Mayor's CATV Advisory Commission."
Fee to the city:	The city will collect 5 percent of the gross revenues each year from each franchise.

The idea of a municipally owned cable system was studied and rejected by Mayor Green, who commented that municipal ownership raised questions of whether the government should control a communications network. Also municipal ownership was prohibited by state law.

Most of the proposed systems were to be joint ventures between a major cable corporation and a local organization. The corporate motivation for entering into these joint ventures was that, in awarding the franchise contract, preference would be given by the city to systems having some local ownership. In a few cases the joint ventures resembled the so-called rent-a-citizen arrangement. For example, the terms of one joint venture would have given the University of Pennsylvania a low-interest loan to purchase a 20-percent share of the ownership in a system to be designed, built, and operated by McLean Hunter, a Canadian firm. In exchange, the university would provide the use of its name and some office facilities. Similar joint ventures were negotiated between other major corporations and several citizens' groups in the city.

Table 7-4
A Summary of the 1979 Philadelphia Cable-Television Proposals

Franchise Area	*Channel Applicant and Corporate Parent*	*Monthly Subscriber Fee for Basic Service*[a] *($)*	*Forward-System Channel Capacity*[b]
1	Telesystems, owned by Times Mirror Corporation	7.00	36
2	JNW Telecommunications, Inc. (a local, minority-owned corporation created to build the system in West Philadelphia)	8.00	36
3	Northwest Philadelphia Cable Communications, Inc. (a subsidiary of Storer Broadcasting, Inc.)	6.00	36
3	Rollins Outdoor Advertising, Inc.	7.95	35
4	American Cablevision of Philadelphia, Inc. (a subsidiary of Time Inc.; 20% locally owned)	7.50	34
4	Comcast Cablevision of Philadelphia, Inc. (owned by Comcast Inc., which is owned by International Equity Co., which is 84% owned by the Sural Corporation)	7.50	34
4	Maclean Hunter Cable TV Limited (a Canadian corporation) and the University of Pennsylvania (which has 20% ownership)	8.00	35
4	Rollins Outdoor Advertising, Inc.	7.00	35
4	Philadelphia Cablevision, Inc. (a subsidiary of Storer Broadcasting, Inc.)	6.95	35
4	Philadelphia Community Cablevision, Inc.	4.00	36
4	Teleprompter of Philadelphia, Inc., (a joint venture of Teleprompter, the largest cable company in the country, and a local consortium with 20% ownership)	6.95	36
4	Ultracom of Philadelphia, owned by AEL, Inc. (based in Lansdale, Pa., and operator of 32 cable systems)	7.50	66

[a]The monthly subscriber fee does not include pay cable (such as HBO or Prims) or the installation fee, $15.00 in most proposals.
[b]The forward-system channel capacity refers to source-to-subscriber distribution; most proposals also included at least four channels of reverse (subscriber-to-source) distribution. A two-way capability is required by the terms of the franchise.

Much of the public debate over cable in Philadelphia focused on Mayor Green's proposal for a single citywide franchise. Four franchise areas had been specified by the terms of the franchise contract. The mayor argued that the single system would be less costly to operate and would solve the problem of providing uniform service and prices to subscribers throughout the city. Although the municipal authorities had attempted to divide the city into four equally profitable franchise areas, most of the bidding attention was directed at franchise area 4 in the Northeast—the area of highest average income. The services proposed for area 4 were somewhat more diverse than had been proposed for the other three franchise areas. If all cable operators were obliged to provide the same level of service at the same prices, all systems might not be equally profitable because of variations in costs and the level of demand.

The city council opposed the mayor's plan of awarding a citywide franchise, arguing that a multiple franchise allowed for competition and ownership diversity. According to this argument, if the city were to award nonexclusive contracts to each of four systems occupying equally profitable franchises, prices could be expected to stay at efficient levels, and innovations in service would be encouraged by competition. It also appeared that only if multiple franchise contracts were awarded would local minority or ethnic groups have an opportunity to share significantly in the ownership of a cable system. A major corporation seemed bound to win a citywide franchise.

Efficient Prices and Design for the Philadelphia System

The examples of efficient pricing and design in chapters 5 and 6 were based on conditions under which eight firms prepared proposals for the Northeast franchise, area 4 in Philadelphia. The conclusions reached in these chapters generally apply to all four franchises. Results of the economic model described in appendix 5B indicate that the municipal authorities had accomplished their goal of dividing the city into four almost equally profitable franchise areas.

The conclusion of the cost-benefit analysis conducted in chapter 6 and summarized in table 7–5 is that the static optimum system design for Philadelphia franchise area 4 would have about 40 channels of capacity and a basic-cable price of $2.40 per month—Design Seven. The four pay-cable services included in Design Seven of the cost-benefit analysis were also typically included in the designs proposed for Philadelphia, and at approximately the same prices. The dynamic optimum, allowing for the expected short-term growth in available services, would have about 80 channels of capacity and a basic-cable price of about $3.50 per month—Design Eight. Substantial savings could be achieved by serving the entire city with a single cable system. Basic cable could be offered at about $1.00 per month, even with an 80-channel capacity—Design Nine.

Most of the private companies bidding in Philadelphia chose a system similar to the static optimum, Design Seven, but including a few locally programmed channels and a separate institutional network dedicated to municipal use. As indicated in table 7–5, adding three locally programmed

Table 7-5
Results of the Cost-Benefit Analysis of Philadelphia

Design	*Efficient Basic-Cable Price (1979 Dollars)*	*Forward Channel Capacity*[a]
Seven: The static optimum for franchise area 4; including four pay-cable and four public-access channels, but no locally programmed channels (as described in chapter 6).	2.40	35–40
Seven: With three additional locally programmed channels.	3.80	35–40
Seven: With three additional locally programmed channels and an institutional network dedicated for municipal use. This design is approximately the same as those designs proposed for Philadelphia in 1979.	6.00	35–40 (and an institutional network)
Eight: The dynamic optimum for franchise area 4; including four pay-cable and four public-access channels, but no locally programmed channels (as described in chapter 6).	3.50	70–80
Nine: The citywide dynamic optimum. A single cable system serving the entire city and offering the same programming as Design Eight (as described in chapter 6).	1.00	70–80
Nine: With three additional locally programmed channels and an institutional network dedicated to municipal use.	2.00	70–80

Note: Pay-cable services included in the cost-benefit analysis are comparable in price and number to the pay-cable services included in the 1979 Philadelphia proposals, listed in table 7-4. The cost-benefit analysis also assumes a modest, not elaborate, investment in public-access equipment.

[a]There will be fewer than 40 channels per cable in the forward system that is used to distribute programming to subscribers if some channels are dedicated to two-way capability or if a few channels at the low end of the spectrum cannot be used because of interference from local television signals. Two 40-channel cables are used to provide 80 channels of system capacity.

channels to Design Seven requires that the monthly subscription price be raised from about $2.40 per month to about $3.80 per month. This price increase can be expected to result in a net loss of consumer's surplus. The loss may be modest, however, because the benefit of locally programmed channels partially offsets the cost of such channels.

In order to support an institutional network for the city, the price of basic cable must be raised from about $3.80 to about $6.00. In effect, the city is imposing a tax of about $2.20 per month on each cable subscriber to pay for the institutional network.

The proposed price of Northwest Philadelphia Communications, Inc., is $6.00 for basic cable in franchise area 3. According to the cost-benefit analysis, this price can be expected to result in a normal profit for a cable system with a few locally programmed channels and an institutional network. In most of the Philadelphia bids, the proposed basic-cable price was in the $6.95–$8.00 range. Allowing for the extensive investment in public-access facilities included in many of these proposals, the expected rate of return on equity is within the range of normal profits—between about 4 and 12 percent in real terms—as estimated in chapter 3. The lower price bid by Philadelphia Community Cable as part of the proposal that is most similar to the static optimum, Design Seven, reflects a more modest investment in public-access facilities, local programming, and the institutional network. The price proposed by Philadelphia Community cable, $3.80, is nearly identical to the $4.00 price suggested by the cost-benefit analysis.

The proposal that most nearly resembles the dynamic optimum, Design Eight, was submitted by Ultracom of Philadelphia. The 66-channel capacity proposed for the forward network, which distributes programming to subscribers, is a few channels less than the capacity suggested by the cost-benefit analysis; Ultracom devoted extensive capacity to two-way and institutional uses. The Ultracom design was an innovative tricable system: one cable was completely dedicated to subscriber use; a second cable was split between subscriber and institutional use; and a third cable was dedicated to institutional use.

To finance the extensive institutional network, Ultracom proposed a $7.50 basic-cable price, $4.00 per month higher than the price suggested by the cost-benefit analysis for Design Eight—a similar design but one that did not include an institutional network.

One firm, Rollins Outdoor Advertising, proposed to combine franchises 3 and 4 into an area that could be served by a single cable system. Firms were, however, discouraged from bidding for all four franchises with a proposal for a citywide system.

The conclusion of the cost-benefit analysis reported in chapter 6 was that a citywide system with about 70–80 channels would be the best design. Although the cost-benefit analysis neglects the adverse effects of reducing competition by allowing only one firm to operate in the city, it indicates that substantial savings could be achieved with a citywide system.

For an 80-channel system without locally programmed channels or an institutional network, the price of a citywide system could be about $1.00 per month, $2.50 lower than if four separate systems were offered. One advantage of the low price is that there would be more subscribers to share in the largely fixed costs of locally programmed channels and an institutional network. In order to support these facilities, the price of a citywide system need be raised only to about $2.00 per month. Raising the price of basic cable, however, does not alter the efficient price of the several pay-cable services, which, because of demand interdependencies, should be set at approximately profit-maximizing levels.

The Price of the Franchise

Before the federally regulated limitation on franchise fees was imposed, some municipalities collected 20, 40, or 50 percent of gross revenues as a franchise fee. In effect, the fee represented the purchase price of the cable franchise, a price charged by the municipality to the cable operator. McDonald's, for example, sells its fast-food franchises under a similar arrangement. When the FCC devised the extensive cable regulations of the mid-1970s, however, it limited the franchise fee to a maximum of 3 percent, subject to review. This regulation is still in force.

The FCC's intention was to restrict fees, the price of the franchise, to a level about equal to the municipality's cost of regulation. To sidestep the intent of the federal limitation, franchising authorities have begun to require payments that are not in the form of fees, such as extensive public-access facilities, grants to develop local programming, and institutional networks. Cable operators are able to raise the price of basic cable high enough to finance these indirect franchise fees because the cable-franchise contract gives them a degree of monopoly market power.

These circumstances add a new dimension to the franchise bidding strategy of private firms, illustrated in figure 7–2. Neglecting the price of the franchise, a single-service cable firm should be expected to select the price P_R associated with output level S_R. For the cost-benefit analysis, P_R represented the Ramsey price that was used in evaluating each of the alternative system designs. The cost of the institutional networks and the other facilities not directly related to subscriber service that the cable operator offers to the municipality as the price of the franchise are not included as part of average total cost in figure 7–2. To finance these facilities, the cable operator must charge price P_F. As a result, fewer subscribers can afford cable service. The difference between S_R and S_F in the figure represents the number of households excluded from cable service by the price increase from P_R to P_F. The shaded area in the figure represents the profits required from the subscriber system to pay for the institutional network and other facilities; the shaded area is therefore the price of the franchise—the monetary value of the facilities the cable operator offers the municipality in exchange for the franchise. In bidding for the franchise, the cable operator must decide what price to offer.

A minimum price is often specified by the terms of the franchise contract, such as a requirement for the private firm to build an institutional network. The maximum a prospective cable operator should be willing to pay

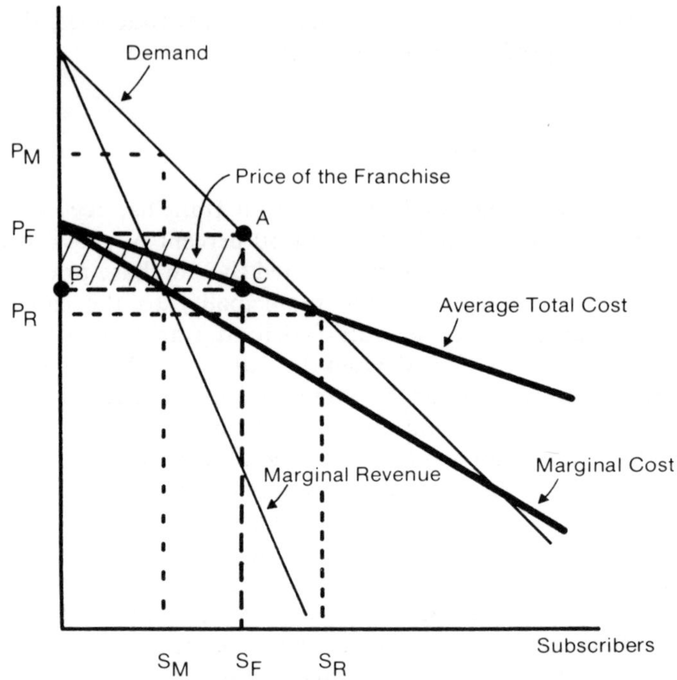

Figure 7–2. The Price of the Franchise, Investment in Facilities for Municipal Use

for the franchise is the profits in excess of average total cost that could be earned by setting the basic-cable price at the monopoly level, P_M in figure 7–2. The goal of the municipal authorities, however, is probably not to extract the highest possible franchise price from the cable operator. Unusually high cable-subscription prices will invoke the wrath of the citizenry. As elected officials, the municipal authorities must be sensitive to the level of cable prices. Under these circumstances, the bidding strategy of the private firms should be to choose the combination of subscription and franchise prices that the municipal authorities will find most attractive.

Conclusions

Allowing for the investment in municipal facilities, which was the price of obtaining the franchise, private firms participating in the 1979 round of franchise bidding for Philadelphia proposed approximately efficient prices for cable services. The franchise-award-winning systems were to be constructed during the period 1979–1983. These were years of explosive growth in the availability of cable services, such as satellite-distributed networks. The cable systems proposed for Philadelphia, however, were not designed to accommodate this growth. Instead firms chose designs suitable for the market of 1979. When the systems became fully operational, in about 1983, they would not have had sufficient channel capacity to accommodate many of the new services that would become available. These design decisions illustrate the difficulty of obtaining the best possible franchise agreement, even over the short run, when the market is changing rapidly.

Based in part on the limited channel capacity of proposed systems, Philadelphia's Mayor Green decided to reject the cable bids and institute a new round of franchising for 1982. The results of the bidding are summarized in table 7-6. Participating firms proposed significantly more channel capacity for the subscriber system, reflecting the increased availability of cable services. The firms also took advantage of the tiering strategy, which offers more varied subscription choices for households, thereby improving economic efficiency.

One drawback to the franchising process is that once cable operators are established as incumbents, they may have a significant advantage in any future round of bidding. In Philadelphia, for example, an incumbent cable operator occupied franchise area 4. The incumbent, Times-Mirror, was the one company to build a system after the 1966 round of franchise bidding. The lack of bids proposed against the incumbent in both the 1979 and 1982 round of bidding reflects the natural-monopoly status of cable systems and the significant barriers to entry, including high investment costs and specialized equipment, for markets in which systems have been established. Perhaps as a result of this lack of competition, the incumbent cable operator, Times-Mirror, proposed the highest basic-cable prices and the lowest channel capacity of any firm participating in the 1982 round of bidding.

A more dramatic example of the potential market power of a franchised cable system is the ability of firms to raise the price of basic cable in order to pay for the institutional network and other facilities that are not subscriber services. Investment in these facilities, which represent the price of the franchise, increased dramatically by 1982 to well over $30 million for Philadelphia. Although the city benefits from having elaborate telecommunications equipment dedicated to municipal use, subscribers must pay substantially higher prices. The higher prices represent a tax levied on cable subscribers in order to finance equipment that would probably not be purchased as part of normal municipal operations. The tax seems unduly burdensome to cable subscribers.

The intent of the federal limitation on franchise fees was to prevent municipalities from using the monopoly market power of franchised

Table 7-6
A Summary of the 1982 Philadelphia Cable-Television Proposals

Franchise Area	*Applicant*	*Basic-Cable Price*	*Channels of Programming*	*Tier*
4	Cablevision	2.00	50	One
		9.50	76	Two
3 and 4	Comcast	2.95	46	One
		8.95	73	Two
3	Cox	5.95	58	One
		11.95	78	Two
2 and 3	Independence	Free	24	One
		3.95	36	Two
		6.95	60	Three
		8.95	88	Four
		12.95	120	Five
2	Intercity	2.00	50	One
		9.50	76	Two
3 and 4	Rollins	1.95	26	One
		3.45	34	Two
		4.95	41	Three
		6.45	53	Four
		7.95	62	Five
1	Times-Mirror	10.90	37	One
		13.85	43	Two
2	Wade Broadcasting	11.95	81	Two

systems to extract excessive payments from cable operators, who pass most of the cost of the payments along to subscribers in the form of higher prices. The franchise fees are intended to offset the cost of regulation. By raising the price of the franchise and taking advantage of monopoly market power, the municipalities restrict the number of households that can afford the communications services of cable. As a result, fewer services will develop—fewer satellite-distributed programs or pay-television networks. By 1983 the effects were apparent. The number of firms participating in franchise negotiations declined and the industry began to loose some of its financial allure.

8 Issues in the Development of Cable Television

In more than 10,000 communities, cable service is available from only one local system. Approximately 30 communities have at some time been served by two overlapping cable systems; in about 20 of those communities, however, one system has eventually gone out of business, leaving the other to operate as a monopolist. The reason for this overwhelmingly monopolist market structure is that the average cost of service is lower when all households in a particular area are served by the same cable system. Additionally, the declining average total costs measured for increasing system length suggest that significant savings can be achieved by allowing a single system to serve a large geographic area. An indication of the potential magnitude of these savings is the relatively low price estimated to be required for a citywide system in Philadelphia—several dollars per month lower than if the city were divided into four smaller franchise areas. To minimize the cost of service, cable operators have been merging small systems into clusters that would serve relatively large geographical market areas. For similar reasons, many small telephone companies merged years ago to form larger operating units.

Like the local telephone company, the electric utility, or the sewage system, a cable system is a natural monopoly. The market power that monopoly status engenders is reason to direct special attention to the issue of efficient prices. If market power is exercised and prices become too high, households in the cable service area are unfairly disadvantaged, and the growth of the cable industry will be dampened. In general, there are two means for achieving efficient prices: competition and regulation.

The case study of Philadelphia supports the premise that when several firms participate in the competitive bidding for a cable franchise, prices are proposed that can be expected to produce approximately normal profits for the firm. Monopoly market power has not been eliminated, however; it is sometimes exercised indirectly by the municipality, which may require extensive investment in public facilities as one of the terms of the franchise contract. In doing so, the municipality uses the monopoly status of the cable system to extract revenue from households in the franchise area. Because federal law prohibits direct payment of this revenue in the form of municipal franchise fees, the municipality receives the payment in the form of facilities.

Ironically, although much of the public concern regarding the abuse of monopoly market power has been directed at the private firm, it is often the municipalities, charged with regulating the industry, that have wielded the market power. Of course, the municipality is wielding this power to obtain facilities that will be used for the common good. At issue is the value of these facilities as opposed to other uses for these economic resources, and the reasonableness of imposing this financial burden on cable subscribers and the cable-television industry.

Competitive bidding for cable franchises appears to be effective, at least in transferring market power from private firms to municipalities. One limitation to the bidding process, however, is that once the franchise has been awarded and a cable operator is established as an incumbent, direct competition is all but removed from the market. To extend the beneficial effects of competitive bidding, the terms of franchise contracts are sometimes written to specify conditions of service, such as price, over the ten- to fifteen-year duration of the franchise. The drawback to this solution is that any change in the market, whether economic or technological, will likely benefit one party to the contract, either the municipality or the private firm, at the expense of the other. For example, higher-than-anticipated inflation or lower-than-anticipated demand could prove to be a financial calamity for a firm that is committed to maintaining prespecified prices for service. Lower inflation or higher-than-anticipated demand would have the reverse effect, resulting in prices that are too high from the perspective of the municipality. To negotiate an ideal long-term contract, both parties must have perfect information about the market. The cable designs submitted to Philadelphia in 1979, which did not anticipate the imminent increased availability of services, demonstrates the difficulty of anticipating future events, even in the short-run.

Municipalities that write long-term franchise contracts will be in a better position to renegotiate if the franchise is nonexclusive. The threat of allowing another operator to establish a competing system in the same market area may be a useful bargaining chip for the municipality. The unwillingness of firms to submit competitive proposals against an incumbent cable operator in Philadelphia, however, suggests that this bargaining chip many not be enough to win the hand.

Other communications technologies are another source of competition that may encourage efficient pricing and design. Historically, the most significant competitor has been broadcast television. Demand for basic cable is more elastic where there are more broadcast stations. Nevertheless, cable has the advantage of more channel capacity and two-way capability. The competitive impact of broadcast television is not sufficient to insure efficient pricing; otherwise municipalities would not be able to take advantage of the market power of franchised cable systems. As more low-power stations are licensed, competition from broadcast television will strengthen. Few industry analysts imagine, however, that these new low-power stations will pose a substantial competitive threat to cable.

The cost characteristics of other technologies such as Satellite Master Antenna Television (SMATV) and direct broadcast satellite (DBS), tend to favor their success in markets not well served by cable. For example, SMATV is particularly cost-effective where the potential subscribers are clumped together, as in hotels, office buildings, or high-rise apartments. The cost advantage of DBS is in a very different market, where low population densities make SMATV and cable prohibitively expensive. The effect of population density on cable and DBS is illustrated in figure 8–1.

Cable is very expensive in markets with low population densities because there are fewer subscribers to share the cost of each mile of distribution plant. With DBS, each subscriber must purchase a small earth station; the distance between subscribers has no impact on this component of cost. Higher population densities do, however, allow more subscribers to share in the fixed cost of the satellite. Because the satellite is a smaller proportion of the total system cost for DBS than the distribution plant is for a cable

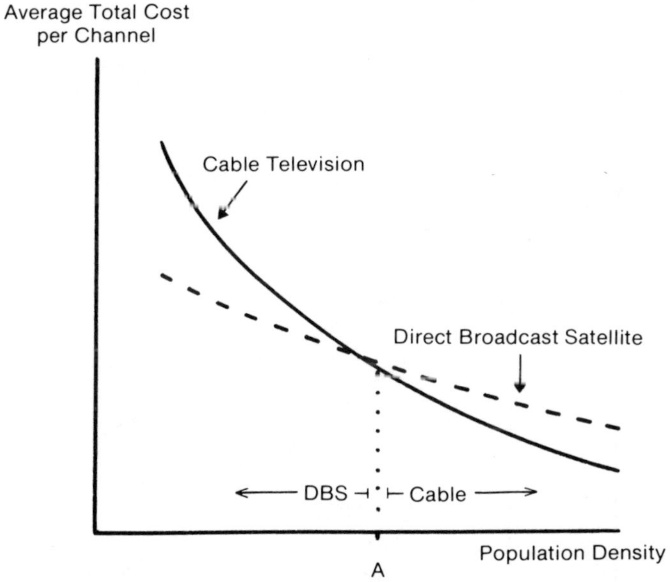

Note: *A* is the market boundary between DBS and cable television. DBS is less expensive for population densities less than A; cable is less expensive for population densities greater than A.

Figure 8–1. A Market Boundary for Cable and Direct Broadcast Satellite Based on Population Density

system, the decline in average total cost associated with higher population densities is less pronounced for DBS than for cable. DBS, then, can be expected to serve rural areas where cable is not economical. Significant competition between the two technologies may occur in areas of moderate population density, perhaps at the perimeter of cable systems. Of course, cable has the advantage of more channel capacity and two-way capability.

Given the expectation that cable systems will continue to operate in most markets as monopolies, some concern has been focused on the concentration of power with respect to programming decisions. Although debate continues in the economic literature over which market structure—monopolistic or competitive—will best respond to consumer preferences, the consensus has favored competition, particularly in view of the almost unlimited channel capacity available with cable techology (Owen, Beebe, and Manning 1974). The major economic drawback in allowing a single firm to make the decisions about the level of programming is, again, that the monopolist can profit by restricting output, as illustrated in figure 8-2. The monopolist, whether it is a private firm or a municipality indirectly exercising monopoly

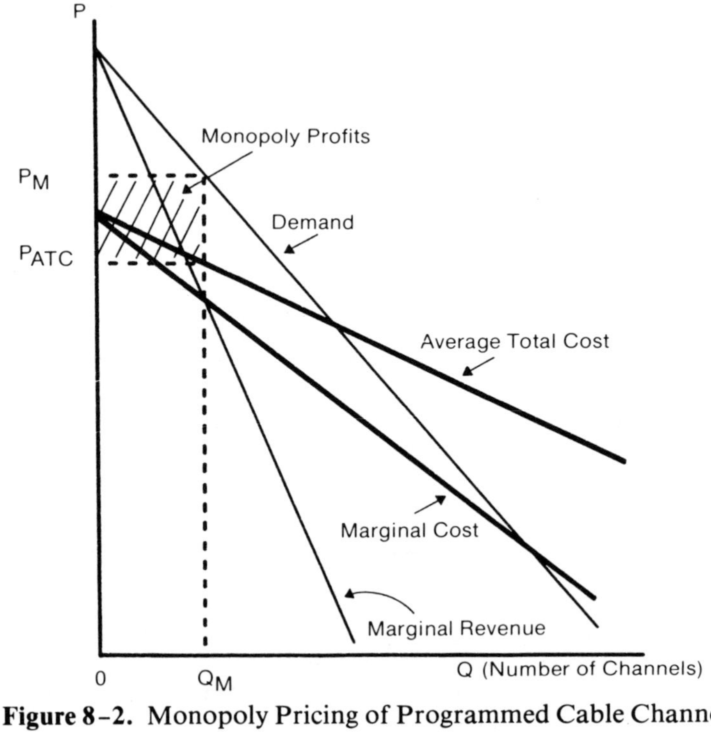

Figure 8-2. Monopoly Pricing of Programmed Cable Channels

power, earns excess profits by charging price P_m for each of Q_m channels of programming. More channels could be programmed, and offered at a lower price, if the firm were restricted by competition or regulation to earning normal profits.

The process of franchising cable television has developed over the years as a combination of competition and regulation. Although municipalities have zealously extracted investments in public facilities and participating firms have not always lived up to their franchise commitments, the process offers an opportunity for price and design competition. It also focuses the regulatory power at the local level, where authorities are more attuned to the particular needs of the community.

The extent to which municipalities should exercise their regulatory jurisdiction will become an increasingly important issue as franchises are awarded and the bidding process can no longer be used to encourage efficient prices and design. Some form of regulation seems to be in order because of the natural-monopoly status of systems, but regulation need not be stifling. Municipalities should act to prevent cable operators from restricting output and charging prices above reasonable levels. To promote innovation and competition in the market for programming services, municipalities can require public and leased access to the cable systems. As a balance to municipal intervention into the market, cable operators and local citizens should discourage excessive investments in public facilities.

The details governing the development of cable television will be negotiated at the local level. The cable operators and municipal authorities have an opportunity to achieve their particular goals by participating in the negotiations. Historically, most cable operators have profited, and a number of municipalities have obtained expensive telecommunications equipment free of cost. Because the franchising process often provides only a limited and largely indirect role in the negotiations for local citizens, however, their needs and goals are sometimes neglected. In many cases they pay prices that are higher than necessary.

Although two-way cable-television systems give citizens a chance to talk back to their television sets, the cable-television market structure restricts their ability to talk back in an economic sense. Perhaps, then, the most important issue for the citizens who are to be served by the industry is how to ensure that cable systems reflect the needs of the community. This issue might make for an interesting program on the public-access channel of the local cable system. As one telecommunications corporate giant has observed, the system is the solution.

Bibliography

Afriat, S.N. *The Price Index.* Cambridge: Cambridge University Press, 1977.

Averch, H., and Johnson, L.L. "Behavior of the Firm under Regulation Constraint." *American Economic Review* 52 (December 1962).

Baumol, W.J. "Contestable Markets: An Uprising in Theory of Industry Structure." *American Economic Review* 72 (March 1982).

Baumol, William J.; Bailey, Elizabeth E.; and Willig, Robert D. "Weak Invisible Hand Theorems on the Sustainability of Multiproduct Natural Monopoly." *American Economic Review* 67 (June 1977)

Bensen, Stanley M. "The Value of Television Time," *Southern Economic Journal* (January 1976).

Bensen, S.M.; Mitchell, B.M.; Noll, R.G.; Owen, B.M.; Park, R.E.; and Rosse, J.N. "Economic Policy Research on Cable Television: Assessing the Costs and Benefits of Cable Deregulation." Washington, D.C.: Office of Telecommunications Policy (OTP), 1976.

Borchardt, Kurt. *Structure and Performance of the U.S. Communications Industry.* Cambridge, Mass.: Harvard University Press, 1970.

Broadcasting-Cable Yearbook: 1980. Washington, D.C.: Broadcasting Publications, 1980.

"The Cable Timetable." *Philadelphia Magazine,* June 1979.

Cable TV Study Committee. *Cable Television in Detroit: A Study in Urban Communications.* Report prepared for Common Council, City of Detroit, May 1972.

Cablevision, 19 May 1980.

Charles River Associates. "Analysis of the Demand for Cable Television." CRA Report no. 178-2, Boston, 1978a.

———. "The Audience-Revenue Relationship for Local Television Stations." Boston, 1978b.

Cogensparger, Lee. "A Computer Simulation of Broadband Cable Distribution Systems." Rensselaer Polytechnic Institute, distributed by the National Technical Information Services. Springfield, Va., 1974.

Comanor, William S., and Mitchel, Bridger. "Community Antenna Television Systems and Local Television Station Audience." *Quarterly Journal of Economics* 80 (1966).

———. "Cable Television and the Impact of Regulation." *Bell Journal of Economics and Management Science* 2 (1971).

Dementz, Harold. "Why Regulate Utilities?" *Journal of Political Science* 11 (April 1968).

Diamond, P.A., and Mirrlees, J.A. "Optimal Taxation and Public Production II: Tax Rules." *American Economic Review* 61 (1971).

Donaldson, Lufkin & Jenrette. "Industry Viewpoint: The Cable Television Industry." Lufkin & Jenrette Securities Corporation, New York, 16 November 1979.

Fisher, Franklin M., and Ferral, Victor. "Community Antenna Television Systems and Local Television Station Audience." *Quarterly Journal of Economics* 80 (1966).

Fuss, Melvyn, and Waverman, Leonard. "Regulation and the Multiproduct Firms: The Case of Telecommunications in Canada." Conference Paper no. 65, unpublished, 1980.

Glejser, H. "A New Test for Heteroskedastic Disturbances." *Journal of the American Statistical Association* 64 (1969).

"The Gold Rush of 1980: Prospecting for Cable Franchises." *Broadcasting Magazine,* March 1980.

Gormely, W.T., Jr. "The Effects of Newspaper-Television Cross-Ownership on News Homogeneity." Chapel Hill, N.C.: Institute for Research in Social Science, 1976.

Haight, Timothy R., ed. *Telecommunications Policy and the Citizen.* New York: Praeger Scientific, 1979.

Hallowell, Mary Louise, ed. *The Cable/Broadband Communications Book.* Washington, D.C.: Communications Press, 1980.

Hicks, J.R. *Value and Capital.* Oxford: Clarendon Press, 1939.

Hotelling, H. "Stability in Competition." *Economic Journal* 34 (March 1929).

Intrilligator, Michael D. *Econometric Models, Techniques and Applications.* Englewood Cliffs, N.J.: Prentice-Hall, 1978.

Keeler, Theodore E. "Railroad Costs, Returns to Scale, and Excess Capacity." *Review of Economics and Statistics* 56, no. 2 (May 1974).

Kelly, Daniel. "Deregulation after Divestiture: The Effect of the AT&T Settlement on Competition." Office of Plans and Policies, Federal Communications Commission, April 1982.

Laskin, Paul I., and Chayes, Obram. "A Brief History of the Issues." In *Control of the Direct Broadcast Satellite: Values in Conflict.* Palo Alto: Aspen Institute, 1974.

LeDuc, Don R. *Cable Television and the FCC: A Crisis in Media Control.* Philadelphia: Temple University Press, 1973.

Levine, Larry S.; Kalba, Konrad K.; and Hochberg, Phillip R., *Taxation, Regionalization, and Pole Attachments: A Comparison of State Cable Television Policies.* Cambridge, Mass.: Harvard Program on Information Resources Policy, Publication p. 78–5, August 1978.

MacAvoy, Paul, ed. *Deregulation of Cable Television.* Washington, D.C.: American Enterprise Institute, 1977.

MacFadden, Daniel. "Conditional Logit Analysis of Qualitative Choice Behavior." In *Frontiers in Econometrics,* ed. P. Zarebka. New York: Academic Press.

Malarkey, Taylor, and Associates. *Philadelphia Cable Television Analysis: Initial Report.* Washington, D.C., 1975.

Marshal, Alfred. *Principles of Economics,* 8th ed. London: MacMillan, 1924.

Mason, William F., project leader. *Urban Cable Systems.* MITRE Corporation, 1972.

Merton, Robert C. "On Estimating the Expected Return on the Market." *Journal of Financial Economics* 8 (1980).

Meyer, John, et al. "The Economics of Competition in the Telecommunications Industry." Boston: Charles River Associates, 1979.

———. *Competition in the Transportation Industries.* Cambridge, Mass.: Harvard University Press, 1976.

Meyers, Stewart C. "The Application of Finance Theory to Public Utility Rate Cases." *Bell Journal of Economics and Management Science* 3, no. 1 (Spring 1972).

Mishan, E.J. *Cost-Benefit Analysis.* New York: Praeger, 1973.

National League of Cities. Report: *Regulating Cable Television.* Washington, D.C., September 1981.

Noll, Roger G.; Peck, Merton J.; McGowan, John J. "Economic Aspects of Television Regulation." Washington, D.C.: Brookings Institution, 1973.

Ohls, J.C. "Marginal Cost Pricing, Investment Theory and CATV." *Journal of Law and Economics* 13 (July 1970).

Owen, B.M. "Newspaper and Television Joint Station Ownership." *Anti-Trust Bulletin* 18 (Winter 1973):787–807.

Owen, Bruce M.; Beebe, Jack H.; and Manning, Willard G., Jr. *Television Economics.* Lexington, Mass.: Lexington Books, D.C. Heath and Company, 1974.

Park, Rolla Edward. *Audience Diversion Due to Cable Television: An Analysis of New Data.* Santa Monica, Calif: Rand Corporation, 1971a.

———. *Prospects for Cable in the 100 Largest Television Markets.* Santa Monica, Calif.: Rand Corporation, 1971b.

———. *The Role of Analysis in Regulatory Decision Making.* Lexington, Mass.: Lexington Books, 1977.

Peacock, A.T., and Rowley, C.K. "Welfare Economics and the Public Regulation of Natural Monopolies." *Journal of Public Economics* 1 (1972).

Pennsylvania General Assembly. "The State of Cable Television in Pennsylvania," September 1974.

Phillips, Almarin. "Ramsey Pricing and Sustainability with Interdependent Demands." *Regulated Industries and Public Enterprise.* (1980).

Posner, Richard A. "Natural Monopoly and Its Regulation." *Stanford Law Review* 21 (February 1969).

———. "The Appropriate Scope of Regulation in the Cable Television

Industry." *Bell Journal of Economics and Management Science* 3, no. 1 (Spring 1972).

Ramsey, Frank. "A Contribution to the Theory of Taxation." *The Economic Journal* March 1927.

RAND Corporation. *Cable Communications in the Daytoniami Valley: Basic Report,* R943 KF/ff (January 1972).

Rees, R. "Second Best Rules for Public Enterprise Pricing." *Economica* (August 1968).

Robichek, Alexander A. "Regulation and Modern Finance Theory." *Journal of Finance* 32, no. 3 (June 1978):693–705.

Samuelson, P.A. *Foundations of Economics.* Cambridge, 1947.

Schaffer, Jan. "Tayoun to Offer Cable Bill." *Philadelphia Inquirer,* 7 May 1981a.

———. "Green to Press for a City Wide Cable TV System." *Philadelphia Inquirer,* 25 June 1981b.

Schink, George A., and Thanawala, Sheela. "The Impacts of Cable TV on Local Station Audience." Wharton Econometric Forecasting Associates, March 1978.

Solomon, Ezra. "Alternative Rate of Return Concepts and Their Implications for Utility Regulation." *Bell Journal of Economics and Management Science* 1, no. 1 (Spring 1970):65–81.

Stigler, G.J. "A Theory of Oligopoly." *Journal of Political Economy* 72 (February 1964).

———. *The Organization of Industry.* Chicago: University of Chicago Press, 1968.

Taylor, John P. "Direct Satellite-to-home Broadcasting: Does COMSAT's Pay-TV Plan Bring it Closer?" *Television Age,*, 19 November 1979.

Teleprompter, Inc. "A Proposal for Cable Television in Philadelphia: Area Four." Unpublished, 1979.

Television Factbook, Washington, D.C.: Television Digest, Inc., 1980.

Telser, L.G. "On the Regulation of Industry: A Note." *Journal of Political Economy* 77 (November–December 1969):937–952.

Train, Kenneth. "Optimal Transit Prices under Increasing Returns to Scale and a Loss Constraint." *Journal of Transport Economics and Policy.* 2 (May 1977):185–194.

U.S. Department of Commerce. *City and County Databook: 1977.* Washington, D.C.: U.S. Government Printing Office, 1978.

U.S. Department of Interior. *The National Atlas.* Washington, D.C.: Geologic Survey, 1970.

U.S. Federal Communications Commission. "The Cable Television Pole Attachments Problem: Assessing the Advisability of a Federal Regulatory Response." Unpublished, August 1977.

U.S. Federal Communications Commission. "Constructing and Operating Major Market Cable Systems." Unpublished, 1977.

———. "Reforming Regulatory Policy for Private Line Telecommunications Services: Implications for Market Performances." Working Paper, Office of Plans and Policy, December 1980.

U.S. FCC, Docket no. 21284 (Economic Inquiry), 71 FCC 2d, 632 (1979a).

U.S. FCC, Docket no. 21284 (Economic Inquiry), 79 FCC, 241 (1979b).

U.S. FCC, Docket no. 20988 (Exclusivity Inquiries), 71 FCC 2d, 951 (1979c).

U.S. v. *Midwest Video,* 406 U.S. 649 (1972).

Vogelsang, Ingo, and Finsinger, Jorge. "A Regulatory Adjustment Process for Optimal Pricing by Multiproduct Firms." *Bell Journal of Economics* (Spring 1979):157–171.

Weinberg, Gary. "Cost Analysis of CATV Components." RMC, Inc. Distributed by the National Technical Information Service, Springfield, Va., 1972.

Williamson, Oliver E. "Franchise Bidding for Natural Monopolies—in General and with Respect to CATV." *Bell Journal of Economics* 7, no. 1 (Spring 1976).

Willig, Robert D. "Consumer's Surplus without Apology," *American Economic Review* (September 1976):589–599.

Willig, Robert D., and Bailey, Elizabeth. "Ramsey Optimal Pricing of Long Distance Telephone Services." In *Pricing in Regulated Industries: Theory and Application,* 1971.

Winch, D.M. *Analytic Welfare Economics.* Middlesex, England: Penguin Books, 1927.

Index

About the Author

G. Kent Webb received the Ph.D, M.B.A., and M.S.E. at the University of Pennsylvania, where he studied economics, finance, and engineering. He is currently manager of the Electronics Industry Econometric Service for Gnostic Concepts, a subsidiary of Data Resources, Incorporated. Formerly, he taught at the University of Pennsylvania and Drexel University, and was a manager of capital planning at CBS.